Infra-red Spectra and Structure of Organic Long-chain Polymers

ARTHUR ELLIOTT, Ph.D. D.Sc.
Reader in Biophysics, Department of Biophysics, King's College, 26–29 Drury Lane, London, W.C.2.

LONDON Edward Arnold (Publishers) Ltd.

First published 1969

SBN: 7131 2214 5

*Printed in Great Britain by
Richard Clay (The Chaucer Press), Ltd.,
Bungay, Suffolk*

Preface

During the last twenty years the investigation of the molecular structure of linear chain polymers has been pursued with the help of many techniques. The interest in polymers is twofold; they are technologically very important, and they constitute a large part of the structure of living organisms. The infra-red spectroscopy of polymers—the determination and interpretation of the frequencies and polarization properties of their infra-red absorption bands—has played a part in this development. It is an ancillary technique as far as the structure of the ordered (crystalline) parts of a polymer is concerned, for where this structure is known it has been determined from the x-ray diffraction pattern, though infra-red spectroscopy has sometimes given guidance. Polymers, however, do not consist of ordered regions only, and infra-red spectroscopy may play a useful part in gaining an insight into the structure of the less-ordered or amorphous regions, which give many polymers their characteristic properties.

This book has been written in the main for those engaged or interested in molecular biology and polymer technology. It may be read by those who have studied in other fields than physics or physical chemistry, and should be suitable for undergraduates in their final year, as well as for postgraduate students.

A. E.

King's College
University of London
1968

Acknowledgements

The Publisher's thanks are due to the publishers and authors for their courtesy in granting permission for the use of copyright material:

Pergamon Press Ltd. for Figs 2.5*a*, *b* and *c*, 2.17. 2.18 reprinted from *A Laboratory Manual of Analytical Methods of Protein Chemistry*, ed. P. Alexander and R. J. Block, Vol. 2 (1961).

Dr. K. D. Parker, The University of Leeds, Astbury Department of Biophysics for Fig. 4.12.

Iliffe Books Ltd. for Table 1.1 (R. G. J. Miller and H. A. Willis) reprinted from *Molecular Spectroscopy*, Pt. 2, ed. G. H. Beaven *et al.*

The Chemical Society of Japan, Tokyo, for Fig. 5.10 reprinted from H. Tadokoro *et al.*, *Bull. Chem. Soc.*, Japan, **34**, 1504 (1961).

Perkin-Elmer Ltd. for Figs 2.3 and 2.4.

Research and Industrial Instruments Company, London, for Fig. 2.6*a* and *b*.

Butterworths Ltd. for Figs. 5.5, 5.6, 5.7, 5.8, and Table 5.1 reprinted from E. M. Bradbury and A. Elliott, *Polymer*, **4**, 47 (1963).

Academic Press Inc., New York for Figs 4.5, 4.6, 4.8–4.10 reprinted from *Synthetic Polypeptides* by C. H. Bamford, A. Elliott and W. E. Hanby (1956); Table 4.2 from E. M. Bradbury, *et al.*, *J. Mol. Biol.*, **5**, 230 (1962); Fig. 5.9 reprinted from R. D. B. Fraser and E. Suzuki, *J. Mol. Biol.*, **14**, 279 (1965); and Fig. 5.16 and Table 3.4 from *Infrared Spectroscopy of High Polymers* by R. Zbinden (1964) to whom thanks are also due.

The Faraday Society for Fig. 4.11 from R. D. B. Fraser's paper in *Faraday Society Discussion*, **9**, 378 (1950) and Fig. 5.11 from J. Mann and H. J. Marrinan's paper in *Trans. Faraday Soc.*, **52**, 481 (1956).

John Wiley & Sons Inc. for Fig. 5.12 reprinted from J. Mann and H. J. Marrinan, *J. Polym. Sci.*, **32**, 357 (1958); Fig. 5.17 reprinted from R. G. J. Miller and H. A. Willis, *J. Polym. Sci.*, **19**, 485 (1956); Figs. 5.13 and 5.14 from E. G. Bendit, *Biopolymers*, **4**, 561 (1956).

The American Chemical Society for Fig. 2.15 reprinted from R. N. Jones, *J. Amer. Chem. Soc.*, **74**, 2681 (1952).

Professor K. Nakanishi and the Nankodo Co. Ltd, Tokyo, for Fig. 5.1 reprinted from *Infrared Absorption Spectroscopy* (1962).

The Editor of Spectrochimica Acta for Figs 2.1 and 2.12

from R. D. B. Fraser and E. Suzuki, *Spectrochimica Acta*, **21**, 615 (1965); Fig. 4.3 and Table 4.1 from E. M. Bradbury and A. Elliott, *Spectrochimica Acta*, **19**, 995 (1963).

The Royal Society for Fig. 5.2 reprinted from A. Elliott, *Proc. Roy. Soc.*, **A226**, 408 (1954) and Fig. 4.7 from E. J. Ambrose and A. Elliott, *Proc. Roy. Soc.*, **A206**, 206 (1951). Fig. 4.9 is in part from E. J. Ambrose and A. Elliott, *Proc. Roy. Soc.*, **A205**, 47, (1951).

The American Institute of Physics for Figs. 3.2, 3.4 and 3.6 reprinted from S. Krimm, C. Y. Liang and G. B. B. M. Sutherland, *J. Chem. Phys.*, **25**, 549 (1956); Fig. 3.3 reprinted from R. G. Snyder, *J. Chem. Phys.*, **27**, 969 (1957); Fig. 5.15 reprinted from H. Tadokora *et al.*, J. Chem. Phys., **42**, 1432 (1965); Figs. 5.3 and 5.4 reprinted from T. Miyazawa *J. Chem. Phys.*, **32**, 1347 (1960).

Contents

CHAPTER 1

General Principles

A INTRODUCTION

The infra-red region of the electromagnetic spectrum extends from a wavelength of about 8000 Å or 0·8 microns towards longer wavelengths or lower frequencies. The energy changes in the molecular vibrations which are excited by this radiation are directly proportional to the frequency, in accordance with the quantum relation

$$E = h\nu \tag{1.1}$$

where h is Planck's constant, ν is the frequency of vibration (in cycles per second) and E the energy change. The experimental observations measure λ, the wavelength of the radiation, which is related to the frequency and the velocity of propagation of the radiation c

$$c = \nu\lambda \tag{1.2}$$

In vacuo, the velocity c is independent of the wavelength, and then the frequency is inversely proportional to the wavelength. It is convenient and usual to give the reciprocal of the wavelength in centimetres, when specifying the position of an absorption or emission band. This quantity is often referred to as the frequency or, more properly, as the 'wave number' and is expressed in cm^{-1}. For the most accurate work it is the number of waves per cm *in vacuo*. Tables exist which convert wavelengths in air to wave numbers *in vacuo*. The vacuum correction (about 0·82 cm^{-1} at 3000 cm^{-1}) is hardly significant compared with other errors of measurement in polymer spectra.

B ORIGINS OF MOLECULAR SPECTRA

Molecular vibrations are made manifest in infra-red and in Raman spectra. The energy changes which result in infra-red bands are usually ones in which the electronic state of the molecules does not change. Changes in vibrational and in rotational states of small molecules produce bands in the infra-red region. Because the energy associated with rotational motion is small, rotation bands occur at low frequencies. In polymers, where there is no free rotation, they do not occur. The necessary condition for absorption or emission of radiation to take place is that the dipole moment of the molecule should change during the vibration. Raman spectra are usually observed with visible radiation, which is re-radiated with changes in frequency which correspond to the vibrational energy levels of the molecule. The condition for a Raman line to appear is that there should be a change in polarizibility during the vibration with which it is associated. Raman spectra of polymers will be referred to only briefly.

Whether or not there is a change in dipole moment during a vibration depends on the symmetry of the system. We will consider briefly the symmetry elements which may be present in a small molecule (the additional elements in polymers will be given later in Chapter 3, where selection rules for polymers are considered).

C SYMMETRY ELEMENTS AND SYMMETRY OPERATIONS

In applying considerations of symmetry to molecules, we consider the atoms (nuclei) as points in a fixed coordinate system. If the system has one or more elements of symmetry, it is possible to transform the (non-vibrating) system in such a way that the positions of the atoms appear to be unchanged after the operation. For instance, if the system has a plane of symmetry, then the operation consists in replacing each atom by its image, considering the plane as a mirror. But since this simply replaces each atom by an identical atom, the system appears to be unchanged. A transformation of this kind is known as a *symmetry operation.*

PLANE OF SYMMETRY (MIRROR PLANE, σ)

If reflection of all the nuclei at a plane produces a configuration indistinguishable from the original, this plane is a plane of symmetry. It is usually designated by σ, with subscripts such as h or v to denote horizontal or vertical. In a more useful form the coordinate axes in which the plane lies ($\sigma(xy)$, etc.) are given. In Fig. 1.1, representing a bent triatomic molecule such as H_2O, there are two mirror or symmetry planes for the non-vibrating molecule.

CENTRE OF SYMMETRY (i)

Here the symmetry operation is reflection or inversion at a centre. A line is drawn from each nucleus through the centre, and produced an equal distance on the other side. If there is an identical nucleus at the end of this line, the molecule has a centre of symmetry.

p-FOLD AXIS OF SYMMETRY (p-FOLD ROTATION AXIS, C_p)

The symmetry operation for C_p is a rotation by $360°/p$ about the axis. Every body has at least a 1-fold rotation axis ($p = 1$). The H_2O molecule in Fig. 1.1 has a 2-fold rotation axis, for a rotation $360°/2$ results in a configuration identical with the original. The operation can be repeated any number of times. The direction of the axis may be specified by adding to the symbol the axis with which it coincides, e.g. $C_2(z)$ in the example in Fig. 1.1.

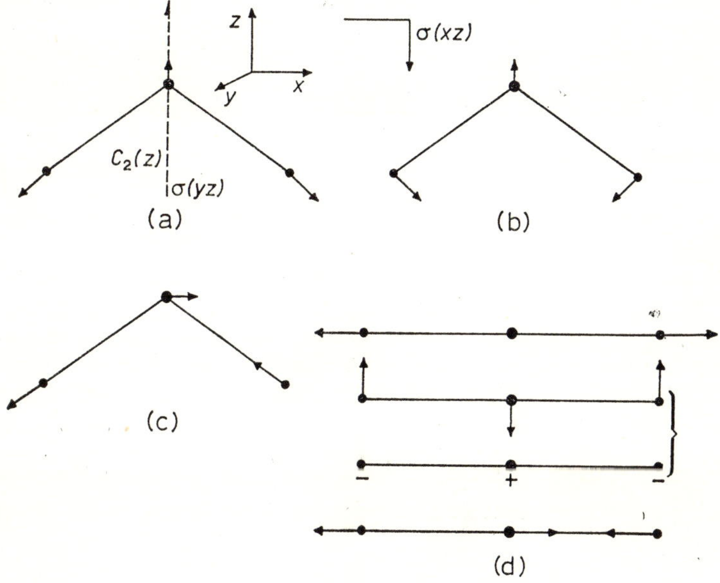

Fig. 1.1 Vibrations in triatomic molecules. (**a**) Symmetric stretching mode and symmetry elements in a bent triatomic molecule; (**b**) deformation mode; (**c**) asymmetric stretching mode; (**d**) vibrations in a linear triatomic molecule. The bracketed pair are degenerate.

p-FOLD ROTATION–REFLECTION AXIS (S_p)

In a molecule with a p-fold rotation–reflection axis the symmetry operation is a rotation of $360°/p$ followed by reflection at a plane perpendicular to this axis. This operation will produce a configuration indistinguishable from the original. The direction of the axis may be specified, e.g. $S_p(x)$.

THE IDENTITY

A symmetry element present in every body is the trivial one E (or I) in which the operation is *to leave the molecule unchanged.* When this is included in a group of symmetry operations, it is found that every symmetry element can be formed by performing, in turn, two of the operations corresponding to the symmetry elements present. This procedure is called *multiplying* the two symmetry elements. Thus,

$$C_2(x) \times C_2(x) = E \tag{1.3}$$
$$C_2(z) \times \sigma(xy) = i, \text{ etc.} \tag{1.4}$$

It may be noted that the operation on the right hand of the product sign is the one to be applied first; the result is not always independent of the order in which the operation is done.

D VIBRATIONS IN SIMPLE MOLECULES

DIATOMIC MOLECULES

The only vibration possible in a diatomic molecule is a motion of the atoms along the internuclear axis. If the molecule is homonuclear (O_2, Cl_2, etc.) it is evident that the centres of positive and negative charges coincide at all phases of the vibration, and the dipole moment remains zero. Vibrations of such molecules do not give rise to infra-red absorption and are said to be *infra-red inactive.* On the other hand, if the atoms are different (HCl, CO, etc.), then the electronic distribution changes during the stretching of the connecting bond, the positive and negative centres of charge do not always coincide and a changing dipole moment is produced. Such active vibrations can interact with electromagnetic radiation, energy being absorbed so that an absorption band appears in the spectrum of the radiation.

Whether we speak in terms of classical or of quantum theory, the frequency of radiation absorbed in a fundamental band is the same as that of the vibrating molecule. On the classical picture, such a band is produced when a non-vibrating molecule absorbs energy of a frequency with which it can vibrate. In the quantum theory, there are no non-vibrating molecules. Each molecule has vibrational energy which is restricted to the values (for the harmonic oscillator)

$$E_v = h\omega c(v + \tfrac{1}{2}) \tag{1.5}$$

where ωc is the frequency of vibration and the quantum number v takes only the values 0, 1, 2, 3, etc. Hence the minimum or zero-point energy corresponding to $v = 0$ is

$$E_0 = \tfrac{1}{2}h\omega c. \tag{1.6}$$

A so-called fundamental transition occurs when energy is abstracted from the radiation beam incident on the molecules, and the quantum number is raised from 0 to 1. At room temperature most of the molecules will be in the state with $v = 0$.

Transitions in which v changes by more than one unit have zero probability if the oscillator is strictly harmonic. For an anharmonic oscillator, however, such transitions have a finite probability, though generally lower than that for fundamental transitions. On the classical model they are ascribed to the weaker components of frequency $2\omega c$, $3\omega c$, etc., present in the anharmonic oscillation whose main frequency is ωc. The corresponding bands in the absorption spectrum are known as *overtone* bands.

POLYATOMIC MOLECULES

The theory of nuclear vibrations in a polyatomic molecule (see, for instance, Herzberg, 1945) shows that the motion can be analysed into a finite number of components of different frequencies, and each component may be considered independently of the others. These components are known as the *normal modes* of the system. The number of normal modes is equal to the number of degrees of freedom of the system. A molecule containing N atoms has, in general, $3N$ degrees of freedom, since each atom requires three coordinates to specify its position with respect to a frame of reference. The molecule as a whole is capable of translation and rotation, and requires $3 + 3 = 6$ coordinates to specify its position, leaving $3N - 6$ degrees of freedom for describing the internal motions of the nuclei. For the special case of a linear molecule (such as O=C=O or H–C≡H) only five coordinates are required to specify the position of the molecule, hence in this case there are $3N - 5$ degrees of freedom for the internal motions. Two of these are of equal frequency, however, and are said to be *degenerate* (an example is given later).

When a molecule is vibrating in one of its normal modes, all atoms move with the same frequency and the Cartesian coordinates of the atoms change harmonically with time. In the case where the motion is not degenerate, all the atoms move in straight lines and go through their equilibrium positions at the same time.

A simple example of normal modes is shown in Fig. 1.1. There are three internal modes. Two result mainly in bond stretching, and are known as the symmetrical and asymmetrical stretching modes, designated by ν_s and ν_a respectively. The third, ν_δ, is one in which the bond angle about the oxygen atom is deformed.

If the bent triatomic molecule of Fig. 1.1 is imagined to straighten out, the stretching modes remain unambiguously defined, but the deformation mode could equally well take place in the plane of the diagram or in any plane containing the (now linear) molecule. The deformation mode can be represented as any combination of two modes of identical frequency, in any two rectangular planes in whose intersection the molecule lies. The mode is described as *doubly degenerate*. It is obvious that the resultant motion of the atoms will in general be an ellipse in this mode.

E GROUP VIBRATIONS

The spectra of a very large number of compounds of known chemical composition have now been recorded, and it has been found that absorption bands which in different materials have rather similar frequencies may often be correlated with chemical groups (often small). This shows that much of the atomic motion in a given

Table 1.1 (Miller and Willis, 1961)

Spectral region	Type of group absorbing in the region	Type of group absorbing on the borderline
I 2–5 μ (5000–2000 cm⁻¹)	X–H stretching X≡X stretching Z–H stretching X=X=X stretching X=X=Y	
II 5–7 μ (2000–1400 cm⁻¹)	X=X stretching (aliphatic) X=X stretching (aromatic rings) X=Y stretching	
		X–H bending X–H wagging
III 7–10 μ (1400–1000 cm⁻¹)	X–X stretching X–Y stretching X=Z stretching Z–H bending	
		X–X stretching X–Y in cyclic systems
IV 10–15 μ (1000–650 cm⁻¹)	X=X–H bending X(H)(H) rocking X–H (aromatic) bending X–Z stretching	
V 15–25 μ (650–400 cm⁻¹)	X–Z bending X–X, X–Y {bending in cyclic or aromatic systems} X–Br stretching X–I stretching	Not often used in analysis

where

		Atomic mass	
H is	hydrogen	1	
X and Y are	carbon	12	Mass group
	nitrogen	14	
	oxygen	16	
	fluorine	19	
Z is	silicon	28	Mass group
	phosphorus	31	
	sulphur	32	
	chlorine	35·5	

normal mode takes place in a small part of the molecule. This is particularly true of modes in which the bond between hydrogen and other atoms undergoes stretching and contraction. Because of the empirical correlations which are now known, it is often possible to identify the origin of an absorption band from its frequency. Since, however, the frequency even for an –X–H mode is not entirely independent of the nature of the other atoms to which X is attached, a further refinement of assignment is sometimes possible.

A few of the more important correlations are given in Table 1.1 (Miller and Willis, 1961), but it must be emphasized that the identification of bands of unknown origin requires great care, and reference should be made to works on the subject (Bellamy, 1958; Nakinishi, 1962).

F TERMINOLOGY USED IN ABSORPTION MEASUREMENTS

When radiation passes through an absorbing medium, the intensity is reduced by a constant fraction per unit of thickness traversed within the medium. If I is the intensity after a distance x has been traversed, the reduction within an element of path dx is

$$dI/I = -\mu \, dx.$$

Hence

$$I = I_0 \exp(-\mu x).$$

The incident light intensity is I_0 and the transmitted intensity for a thickness x is I. The constant μ is the *absorption coefficient.* Evidently,

$$\mu = 1/x \ln (I_0/I)$$

A related quantity, the *extinction coefficient* K, is defined as

$$K = 1/x \log_{10} (I_0/I).$$

These definitions refer to intensity loss within the medium. When radiation is transmitted through a sheet of absorbing material, there are in addition reflection losses at the external surfaces. When the transmitted intensity includes reflection losses, the quantities I/I_0 and log (I_0/I) are called the *fractional transmission* and the *optical density*, respectively.

If reflection losses have been eliminated or allowed for in determining I, the corresponding functions are known as *transmittance* and *absorbance.* When a double-beam instrument is used to make measurements with solutions in cells, it is customary to put a similar cell containing the solvent in the reference beam. This corrects for reflection losses as well as for absorption by the solvent, and the quantity measured is then the transmittance or the absorbance. The charts for double-beam spectrometers are usually scaled in units of these quantities. When measurements are made with solid polymers, the insertion of a transparent window in the reference beam does not completely correct for reflection loss, for the refractive index of the polymer within a strong absorption band (and hence the reflection loss) may be much greater than that of a transparent window. The correction is difficult to make satisfactorily (see Robinson and Price, 1953; Abbott and Elliott, 1956). Because of these considerations, we prefer to use the terms fractional transmission and optical density when it is these quantities which are the ones measured (as is almost always the case with solid polymers).

CHAPTER 2

Experimental Methods

In this chapter the experimental techniques applicable to the spectroscopy of polymers are considered. For a detailed consideration of spectrometer design and performance the reader may refer to such works as Bauman (1962) and Potts(1963). A general account of the spectroscopy of industrial polymers is to be found in a book by Henniker (1967) and a chapter by Fraser (1960) deals more particularly with proteins.

A SPECTROMETER

For recording the infra-red spectrum of polymers which are in the form of thin sheets, the commercially available spectrometers will generally be suitable. Such spectrometers are designed for specimens (often cells containing liquids) several millimetres wide and perhaps 2 cm high. They are almost always of the double-beam type, in which the radiation from a filament or globar is divided into two beams. One beam serves as a reference, and the other passes through the specimen. By means of an oscillating or rotating mirror or other device the radiation is sent alternately through the specimen and over a fixed path in which may be placed a blank cell (i.e. one which does not contain the substance to be measured). Each beam in turn passes into the spectrometer via the entrance slit. It is then dispersed by the prism or grating of the spectrometer, and a narrow region of the spectrum is transmitted by the exit slit, to be concentrated on the detector (usually a thermocouple). In this way a succession of signals is generated, corresponding to the energy transmitted in turn by the specimen (I) and by the blank cell. If the system is arranged so that the energy transmitted by the blank cell is equal to the energy falling on the specimen (I_0), then evidently the ratio of successive signals I/I_0 is equal to the fraction transmitted by the specimen. By means of an optical–mechanical arrangement the recorder is made to record the ratio I/I_0. The wavelength transmitted by the exit slit is continuously varied (usually by means of a rotating mirror in a prism instrument, or by rotating the grating if there is one) and the ratio I/I_0 is recorded over the desired range of wavelengths. The optical arrangement of one type of double-beam spectrometer is shown in Fig. 2.1. These instruments, if in proper

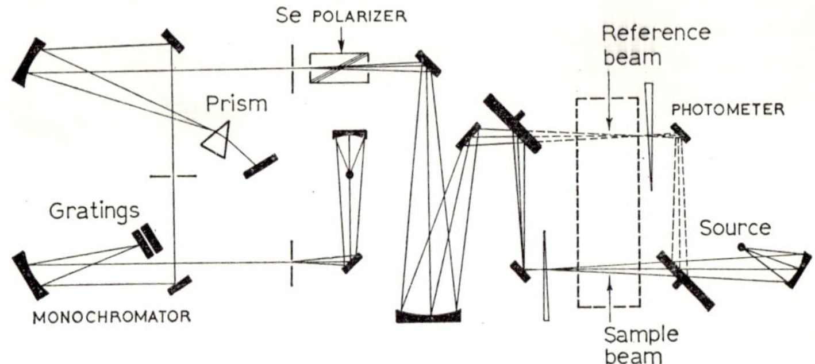

Fig. 2.1 Optical arrangement in a double-beam (Beckman IR9) spectrometer, showing the position of the polarizer which has been added. (Fraser and Suzuki, 1965a)

adjustment, eliminate the absorption bands of atmospheric water and carbon dioxide which would appear in a single-beam record, for the intensity of each beam is reduced by the same fraction and so the ratio I/I_0 is not affected.

In all spectrometers the spectrum is recorded on a chart. Such charts commonly have vertical divisions with the corresponding wavelengths or alternatively wave numbers printed on them. The horizontal lines may be scaled in units of transmittance or of absorbance. It may also be possible to take the output of the instrument to a separate recorder of the strip chart kind, which allows the wave-number scale to be altered by changing the chart speed. In such a case, however, some means for calibrating this scale must be provided. It is usually done by a system of contacts on the periphery of a wheel which is geared to the mechanism for changing the wave number. Through these contacts, a series of impulses is given to the pen, so the spectrum carries a series of short, sharp peaks which can be related to the wave number (see below: Section G, 1). It is desirable to be able to run through a spectrum a second time without losing the wavelength calibration, particularly if polarized radiation is to be used.

Recorders are available which plot not the ratio I/I_0 but $\log_{10} I_0/I$, that is, the optical density. For many purposes the optical density is a more convenient function than the fractional transmission, since optical densities are additive. The replotting of a record of transmission to one of optical density by measurement and hand plotting is tedious, but the labour may be reduced by various devices (see section G, 3).

PRISMS AND GRATINGS

The dispersion of a suitable prism is generally sufficient to resolve the detail in a polymer spectrum, but it is necessary to use different prisms in different spectral regions, in order to get at the same time sufficient dispersion and transparency in the prism. For most purposes, two prisms are adequate. The region from about 3700 to 2000 cm^{-1} can be covered with either a LiF or a CaF prism, and a NaCl prism may be used from 2000 to 600 cm^{-1}. The dispersion of a prism instrument may be increased by sending the radiation through the prism twice ('double-pass') as described by Walsh (1952).

Sodium chloride prisms are hygroscopic, and their polished surfaces become fogged in a damp atmosphere. For this reason, a desiccating agent (renewed as required) should be kept in the spectrometer. Activated alumina or calcium sulphate may be used. It is also good practice to keep the internal temperature of the spectrometer a few degrees higher than that of the air outside the spectrometer, for this will lower the relative humidity, which is the significant factor. The internal temperature is usually maintained constant by a thermostat.

Diffraction gratings have come increasingly into use in spectrometers instead of prisms, and at least one firm has ceased to manufacture infra-red spectrometers with prisms as the main dispersing element. Gratings have much higher resolving powers than prisms, and since reflecting gratings are used, the problem of low transmission does not arise. Gratings have rulings cut at a chosen angle ('blazed') to throw most of their energy into one order, with a consequent gain in intensity. Because they can be made to give spectra of higher angular dispersion than can be obtained from prisms, wider slits may be used for a given resolving power. With absorption spectra which are recorded against a background of continuous radiation, the energy flux falling on the detector is proportional to the square of the slit width. In consequence, if two instruments with different angular dispersions have their slits adjusted to give equal resolving powers, the instrument which has the higher dispersion will (other factors being equal) give the higher output.

Diffraction gratings do, however, have some disadvantages. Different orders overlap, and either filters or a small prism must be used to remove the unwanted radiation. As explained in the section on procedures, the reflecting power of a grating depends on the direction of the electric vector of the radiation, and this is a serious complication when polarized radiation is used. A third possible disadvantage occurs when the specimen is available only as a narrow strip (for instance, a cut section of a fibre), for a wide slit requires a wider specimen than a narrow slit does. The high sensitivity of modern grating instruments may make it possible to obtain satisfactory spectra even in this case, provided that the specimen is masked by a slit so that only radiation passing through the specimen is transmitted. A similar slit would be inserted in the reference beam.

B ILLUMINATING SYSTEMS

For recording the spectra of small specimens (frequently necessary with polymers), special illuminating and magnifying systems have been used. Most of these have aluminized curved mirrors instead of lenses. They are designed to form a magnified image of the specimen on the entrance slit, with an angular aperture of the converging rays

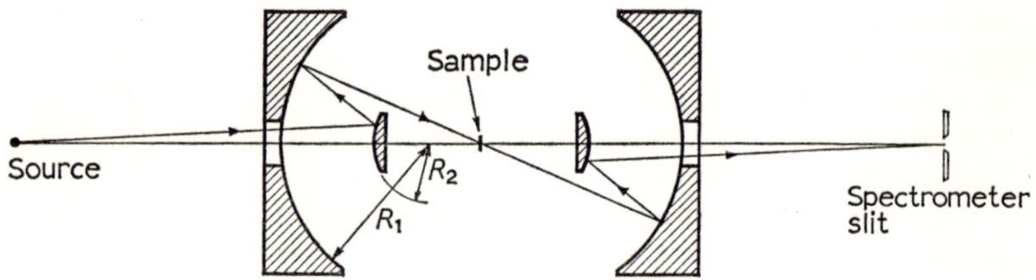

Fig. 2.2 Reflecting microscope. Dimensions in Table 2.1.

sufficient to fill the aperture of the collimator mirror after passage through the slit. This requires the specimen to be illuminated by means of a condensing system which forms a reduced image of the source on the specimen (see Figs. 2.2–2.4). Usually condenser and magnifier are identical. Such systems constitute reflecting microscopes. Some of the early forms have been described by Fraser (1960) and by Elliott (1959).

One system which has been much used is based on the concentric mirror design of Schwarzschild. Norris (1954) has described three slightly different versions of this which he calls A, B and C. Fraser (1953a) has used design A, and the author and Dr Bradbury have used the B type (see Bradbury and Ford, 1966), the dimensions in this case being scaled up. Table 2.1 gives the dimensions and Fig. 2.2 shows the arrangement. The dimensions of the convex mirror are chosen so that the beam just fills that part of the spectrometer collimator which is used (this is always less than the whole aperture). Since prisms and gratings are rectangular, the convex mirror is also made rectangular in order to minimize the central obscuration which it produces. The adjustment of reflecting microscopes of the type shown in Fig. 2.2 has been described by Norris, Seeds and Wilkins (1951).

Table 2.1 Dimensions and optical constants of a reflecting microscope (Bradbury and Ford, 1966)

Radius of curvature of concave mirror R_1	6·7 cm
Radius of curvature of convex mirror R_2	2·57 cm
Diameter of concave mirror	11·30 cm
Diameter of convex mirror	2·57 cm
Numerical aperture (NA)	0·64
Focal length	2·08 cm
Working distance	4·83 cm
Central obstruction	45%
Magnification	5·7
Mirror separation	4·1 cm
Tube length	11·9 cm

The magnification which can be employed when a reflecting microscope is used in the way shown in Fig. 2.2 is restricted by the numerical apertures of the microscope and spectrometer. If these are, respectively, NA_0 and NA_s, then the collimator will be just filled with radiation when

$$NA_0/NA_s = m$$

where m is the magnification produced by the microscope. If higher magnification is used, the outer parts of the collimator will not be filled. The loss in sensitivity may be very great, since the radiation from the central portion of the collimator mirror may be obscured by the detector and its housing. According to Norris, reflecting microscopes with spherical surfaces and numerical apertures up to 0·8 are satisfactory when used with infra-red spectrometers. However, if polarized radiation is to be used it is better to restrict the numerical aperture to about 0·6, otherwise the angles of incidence of the radiation on the specimen are rather high, and this introduces errors in the measurement of dichroic ratios. With a numerical aperture of 0·8, the magnification possible with the usual spectrometer design is about 11.

In the arrangement shown in Fig. 2.2 the undispersed beam falls on the specimen, whose temperature is raised by absorption of radiation. The temperature rise may be greatly reduced by interchanging the source and detector. The dispersed beam coming from the spectrometer which is concentrated on the specimen by the reflecting microscope objective has a much smaller energy content than the whole beam, and specimen heating is greatly reduced. This arrangement has been used by Elliott (1952) for observing the infra-red spectrum of single crystals of ribonuclease, which would have been affected by the heat absorbed from the undispersed beam. A similar disposition has been used by Coates, Offner and Siegler (1953) in the Perkin–Elmer microspectrometer shown in Fig. 2.3.

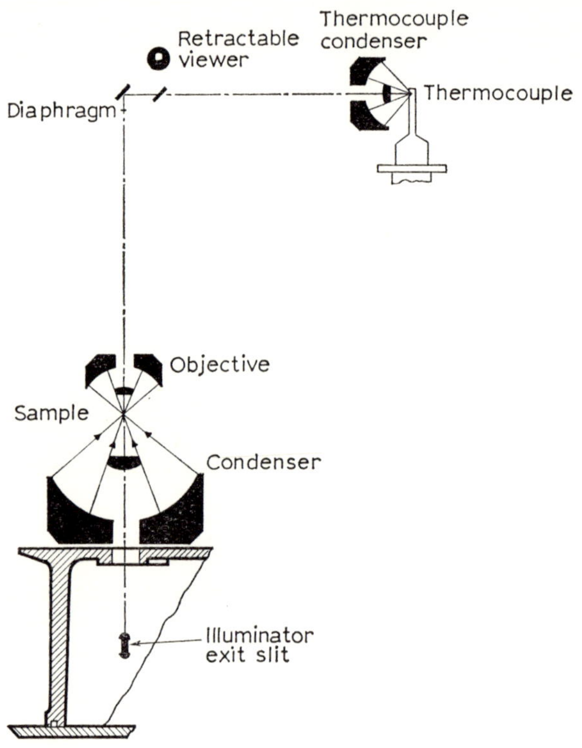

Fig. 2.3 Perkin–Elmer microspectrometer. (Coates, Offner and Siegler, 1953)

A refracting micro-illuminator or beam condenser using potassium bromide lenses which can be used with a number of commercial double-beam instruments is made by Research and Industrial Instruments Ltd. of London and by the Limit Corporation in the U.S. The beam condenser consists of two lenses between which the sample is placed, and permits a specimen reduction to one-quarter. The reference beam should be suitably attenuated to compensate for the reduction in intensity of the sample beam.

A micro-illuminator produced by the Perkin–Elmer Corporation which uses off-axis ellipsoidal mirrors is shown in Fig. 2.4. This gives a magnification of $\times 6$. The unit may be put in the well of a double-beam instrument as shown. The air path of the sample beam is increased by using this arrangement, and atmospheric absorption bands will appear in the spectrum unless some compensating arrangement is used. A similar unit placed in the reference beam would restore the compensation. It should be noted than an ellipsoidal system produces large aberrations in that part of the image which is not on the axis, and small specimens should be mounted over a slit.

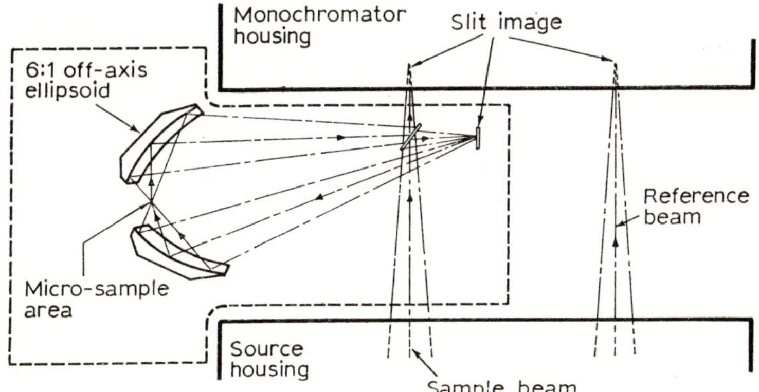

Fig. 2.4 Off-axis ellipsoids giving a magnification of $\times 6$. (Perkin–Elmer Corporation)

The use of a micro-illuminator or reflecting microscope with a double-beam spectrometer not designed for this combination produces difficulties. One way of retaining some of the advantages of the double-beam principle with a single beam is to move the specimen in and out of the beam, recording alternately I and I_0. Elliott (1959) has described a simple specimen-rocking arrangement which returns the specimen to the same position at the end of each cycle with high precision. The same principle has been developed by Ford, Price, Seeds and Wilkinson (1958) who have described a spectrometer which records the ratio I/I_0 throughout the spectrum. More recently an improved version has been described by Bradbury and Ford (1966). The reflecting microscope used is shown in Fig. 2.2. Examples of the spectra obtained are shown in various papers by Bradbury and Elliott (see Figs. 5.5 and 5.6 in this book).

C POLARIZERS FOR INFRA-RED RADIATION

Many examples of the use of polarized radiation in infra-red spectroscopy will be given in later chapters. The most convenient and generally useful polarizer for infra-red radiation is the transmission type. Transmission polarizers have been used occasionally in the visible region for a very long time; they consist of a pile of thin transparent plates inclined at the polarizing (Brewster) angle to the beam, so that the angle of incidence θ on the front face of each plate is given by

$$n = \tan \theta$$

where n is the refractive index for the wavelength of the light.

The first infra-red polarizer of this type had thin films of selenium (Elliott and Ambrose, 1947; Elliott, Ambrose and Temple, 1948a).

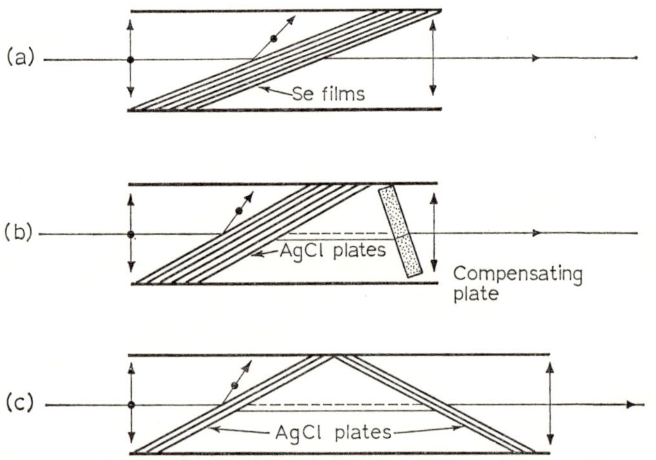

Fig. 2.5 (**a**) The optical arrangement in a selenium transmission polarizer; (**b**) and (**c**) methods of compensating for the image displacement introduced by silver chloride polarizers. (Fraser, 1960)

Selenium in the amorphous form is very transparent in the infra-red region, from the visible limit to 18 μ and again beyond 21 μ (Gebbie and Cannon, 1952). Films of a few microns thickness can be made by evaporating selenium *in vacuo* on to a temporary substrate, which is later dissolved off. A simplified technique has more recently been described by Bradbury and Elliott (1962). The temporary support is polystyrene film about 15 μ thick. This material is dimensionally stable, and a supply of the coated film remains good in store. The selenium coating is cemented to thin metal frames with a protein glue, and the base is dissolved off in carbon tetrachloride (rapidly penetrating solvents cause disruption of the film and are not suitable). The selenium films on their frames are mounted as shown in Fig. 2.5a, six being a suitable number. At each reflection (internal as well as external) a fraction of the radiation whose **E** vector is perpendicular to the plane of incidence is reflected away, and the transmitted beam from six films consists almost wholly of vibrations in the plane of incidence. If the intensities of the components of the original beam vibrating in planes parallel and perpendicular to the plane of incidence are I_π and I_σ respectively, the fractional polarization is

$$\frac{I_\pi - I_\sigma}{I_\pi + I_\sigma}.$$

The value of this fraction is higher the greater the number of plates and the higher the refractive index. For six plates of selenium (whose refractive index is about 2·5 in the infra-red region) the fractional polarization is not less than 0·98, and the fraction of the original beam transmitted is about 0·44 (Elliott, Ambrose and Temple, 1948a). Conn and Eaton (1954) have found that the fractional polarization of selenium polarizers can be extremely high. It is important to avoid too uniform a film thickness in the polarizer, otherwise interference effects can be troublesome (Abbott and Elliott, 1956). Three different film thicknesses, chosen so that their interference bands do not overlap, are recommended. The selenium polarizer, having very thin films, can be introduced into a converging beam of radiation without affecting the focus of the beam. The plane of polarization is rotated by rotating the polarizer, and again because of the small thickness, the image is not appreciably rotated. Selenium films are somewhat fragile, but with reasonable care the polarizers remain serviceable for some years.

Silver chloride ($n = 2{\cdot}0$) can be rolled into thin sheets and has been used for making transmission polarizers (Newman and Halford, 1948; Wright, 1948). Because of the lower refractive index, the fractional polarization is lower than is produced with the same number of selenium films, and is 0·92 for a pile of six films. Silver chloride is robust, but has not been made so thin that the beam displacement and change in focus is negligible. Various methods for compensating the effect of refraction have been proposed; two of these are shown in Fig. 2.5. In (*b*) a transparent plate or wedge is used (as in the polarizer made by Perkin–Elmer) and in (*c*) the first three plates produce a deviation which is compensated by the last three (Makas and Shurcliff, 1955). This last arrangement suffers from the rather serious disadvantage that it reduces the angular aperture. Makas and Shurcliff recommend putting the plates on either side of a focal point, such as the entrance slit of the spectrometer, in order to avoid reducing the angular aperture of the beam.

Silver chloride should be stored in the dark, and should not be mounted in contact with base metals.

A new type of infra-red polarizer has recently become commercially available (Perkin–Elmer Corporation). It consists of a grating formed on a substrate of silver chloride by uniformly spaced parallel strips of evaporated gold (2880 lines per mm). This acts as an efficient polarizer in the region 2·5–30 μ. It is used with the radiation at normal incidence, and produces no lateral displacement. Little space is taken up by the polarizer, which is only 3·5 mm thick. The following performance figures are taken from graphs in the manufacturer's catalogue: the transmission of incident plane polarized radiation is 60% at 4 μ, and 75% between 8 and 20 μ (30% and $37\frac{1}{2}$% transmission of unpolarized radiation respectively). The fractional polarization as defined above is 0·96 at 2 μ, 0·993 at 6 μ and at wavelengths greater than 8 μ it is constant at 0·996.

The performance is very good, and the polarizer is robust and convenient. It may be obtained with a clear aperture of 22 mm.

D TOTAL REFLECTION METHODS

When radiation is totally internally reflected at the boundary of a dielectric, there is penetration of the waves into the less dense medium, amounting to something less than the wavelength of the radiation. If the less dense medium has absorption at this wavelength, energy will be removed from the beam. Hence it is possible to observe the absorption spectrum of a sample by putting it in optical contact with a transparent material of higher refractive index and analysing the beam totally reflected at the interface. The method has been developed by Fahrenfort (1961). It is useful for obtaining an absorption spectrum from a specimen which is too thick, or which is in the form of a layer on an opaque substrate (for example, a layer of paint or varnish on metal). The dielectric should have a refractive index 2·0 or higher for use with most organic substances. A mixed crystal of thallium bromide and thallium iodide (known as KRS5) is often used, and silver chloride and germanium are also suitable. KRS5 is attacked somewhat by water. Apparatus for attenuated total reflection (ATR) spectra is available commercially, and is made to fit most commercial spectrometers of American and British design. Figure 2.6a shows one arrangement made by the Research and Industrial Instrument Company Ltd. of London. Radiation is reflected from the fixed plane mirror M1 on to the toroidal mirror T1 so that it is focused outside the hemicylinder made of material of high refractive index. An approximately parallel beam is produced within the hemicylinder, so ensuring that the angle of incidence on the flat face is nearly the same for all rays. The radiation is here internally reflected, and if a specimen is put in optical contact with this flat face, the absorption spectrum of the specimen will appear when the reflected beam is examined. The reflected beam is returned by the mirrors T2 and M2 to the entrance slit of the spectrometer. The angle of incidence on the back face of the hemicylinder can be varied, and the mirrors T2 and M2 are made to rotate appropriately so as to keep the focus of the beam on the spectrometer slit.

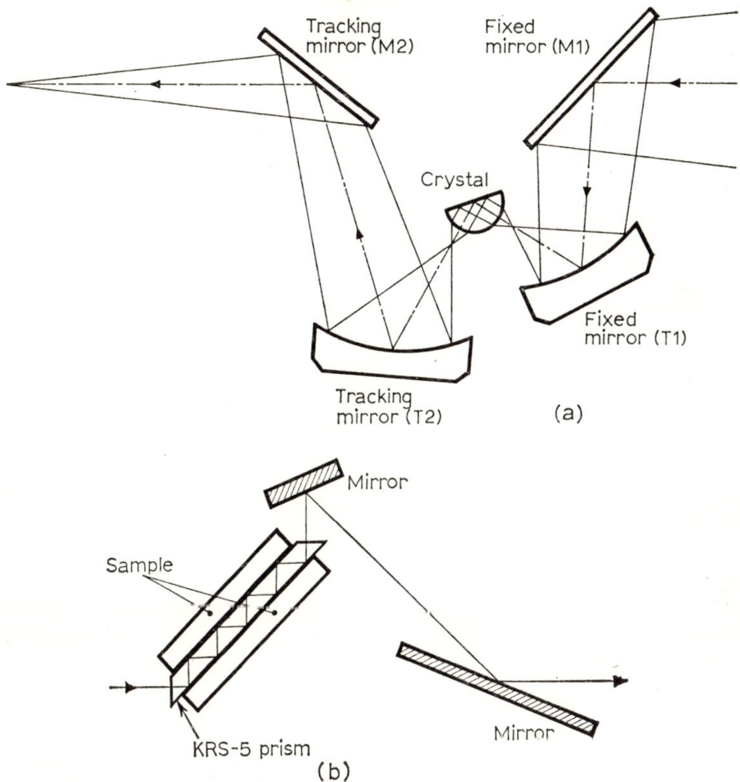

Fig. 2.6 Arrangement for observing attenuated total reflection spectra: (**a**) micro-beam arrangement. Reflection occurs at the flat face of the hemicylindrical crystal; (**b**) prism giving multiple reflection. (Research and Industrial Instruments Ltd.)

It is possible to record weaker absorption bands by reflecting the beam several times. Figure 2.6b shows how this may be done with a suitably shaped prism. It is found, in practice, that the prism with flat entry and exit surfaces is satisfactory. Fahrenfort has used carbon disulphide as an immersion liquid to make optical contact, and has found that contact is often maintained after the liquid has evaporated. Although ATR spectra are very similar to the corresponding absorption spectrum, small differences do exist and some bands do not occur at exactly the same frequency in both spectra. More information may be found in a book by Herrick (1967).

E SPECIMEN PREPARATION

Solid polymer specimens intended for spectroscopic work in the region 3–15 μ should be a few microns thick, and if they contain polar groups such as C=O, NH and OH linkages they may have to be only about 1 μ thick if the strongest bands are to be recorded satisfactorily. In general, thicker specimens will be required for the longer wavelengths. If quantitative measurements are required, the specimen thickness should be fairly uniform. On the other hand, very uniform thin films with optically reflecting surfaces will give spectra crossed by interference bands and accurate measurements of transmittance or of optical density are then almost impossible. It is usually possible to avoid interference effects, either by roughening the surfaces of the specimen or by mounting it in optical contact with a much thicker plate of transparent material such as barium fluoride, sodium chloride, silver chloride, etc.

ISOTROPIC FILMS

Polymer films are hardly ever isotropic in the full sense of having identical properties in all directions, for there is usually a tendency for long-chain molecules to lie in the plane of the film (uniplanar orientation), but they may be made so that the properties in all directions within this plane are the same. In general, three methods are available.

(*a*) Evaporation of the solvent from a solution of the polymer spread on a horizontal surface.
(*b*) Precipitation of the polymer from a solution spread on a substrate, which is immersed in a liquid in which the polymer is insoluble but which is miscible with the solvent.
(*c*) Melting the polymer between two flat plates to give a thin film which is subsequently separated from the plates.

Those who are familiar with the terminology used in textile manufacture will recognize the methods as similar to 'dry', 'wet' and 'melt' spinning of fibres. Some practical details will be given.

(*a*) *Dry casting* Large pieces of film may be made by using the arrangement shown in Fig. 2.7. A piece of plate glass is cleaned thoroughly and treated with a preparation such as a solution of dichlordimethylsilane in carbon tetrachloride (Repelkote) by swabbing. When the solvent has evaporated, the plate is washed in water and it will be found to have a water-repelling surface from which most polymers can easily be separated. The application should be done in a fume cupboard or out of doors. Two parallel wires are clamped at one end of the plate and are kept taut by weights or other means. (It is convenient to use copper wire which has been stretched slightly to remove kinks.) The plate is levelled carefully, and a pool of polymer solution is poured on the plate between the wires and is spread to a uniform thickness by means of a straight-edge resting on the wires. It is then allowed to dry. It is usually easy to cut the film to the required size with the point of a razor-blade or scalpel and then to peel it off.

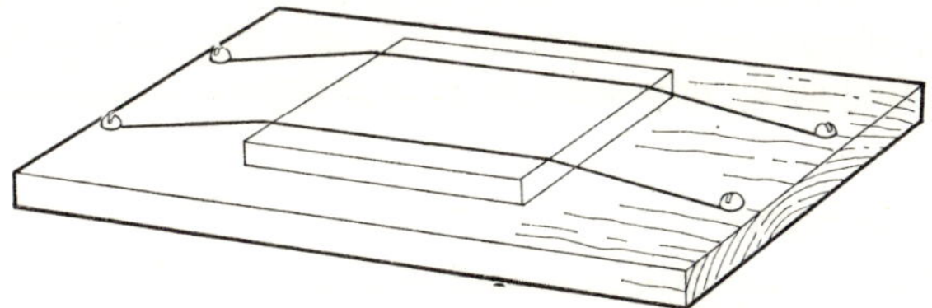

Fig. 2.7 Arrangement for dry casting a polymer film.

If little material is available, it may be better to apply the solution to a plate of transparent material (horizontal) and to leave the subsequently dried film on the substrate. Greater uniformity of thickness will be obtained if the drying is done slowly, say by covering the coated plate with a small inverted beaker so that the solvent can escape slowly.

Some solvents are tenaciously held by polymers even when they are apparently dry, and heat treatment (preferably in a vacuum) may be needed to remove them.

(*b*) *Wet casting* Wet casting is useful when the solvent contains material which is not easily evaporable (for instance, inorganic salts). It may also be useful if it is desired to reproduce, in a film, some of the conditions under which wet-spun fibres are made. The polymer solution is spread on a substrate and gently lowered into the precipitating liquid. The physical consistency of the resulting film varies greatly with the nature of the precipitant. It will generally be found that coherent films are more often produced by precipitants which have similar solvent properties to the polymer solvent than by those which are very different. It may be useful to try a range of precipitants chosen from a homologous series with, for instance, increasing fractions of hydrocarbon groups. Usually, polymer films which have been cast in the way described will be in the form of swollen gels, which should be washed to remove the solvent before they are dried. Such gels will usually form uniplanar oriented films when dry.

(*c*) *Melt casting* Polymers which melt at a suitable temperature without decomposing can be formed into films by putting some pieces between two polished plates of Pyrex glass and heating the whole under pressure. It may be advantageous, if very thin films are required, to select a piece (free from air bubbles) from the first attempt, and to remelt it. Quick cooling will usually give a sample with little or no crystallinity.

MULLS AND PRESSED-DISC TECHNIQUES

Specimens in the form of small particles dispersed in liquid paraffin or in a disc of potassium chloride are often used when the spectrum of powders is required. The difficulty with polymers is that it is frequently impossible to reduce them to a sufficiently fine powder.

ORIENTED SPECIMENS

To make use of polarized radiation, at least partial orientation of the polymer chains along a common axis in a specimen is needed. This is usually achieved by mechanical working such as rolling or drawing. Films which are to be drawn should be of uniform thickness and width, and should have edges which have been cut cleanly with a sharp knife, razor-blade or scalpel. Any tear in the edges will develop further under the drawing action. Some polymers may be drawn without difficulty at room temperature; others may break when so treated. In such cases, drawing at a higher temperature may be tried. Another technique which may assist the drawing is the use of a softening or plasticizing agent. Wet steam may be effective. Sometimes a small quantity of a solvent for the polymer may be introduced to act as plasticizer.

An alternative to drawing is mechanical rolling between the cylindrical rollers of a small jeweller's mill. It is difficult to carry out this operation on the very thin films required for infra-red spectroscopy, but it is easier if the films are cast on a deformable substrate which can be rolled thin along with the film. Lead and cadmium have been used as temporary supports for rolling. Silver chloride sheet has mechanical properties which are similar to those of lead, and since it is transparent the polymer, in many cases, need not be removed. If required, the free film may be obtained by dissolving away the silver chloride in aqueous sodium thiosulphate solution (the photographer's 'hypo').

Many polymers in solution become oriented to some extent if the solution is sheared, and show the phenomenon of flow-birefringence. If such solutions are continually sheared during the drying process, some orientation of the molecules in the dried film is to be expected. With some materials the degree of orientation is very high. In such cases the molecules form straight rigid rods when in solution and even at rather low concentration (10% or sometimes less) these solutions are liquid crystals. They are spontaneously birefringent, and it is quite difficult to prepare isotropic films from them. Examples are: tobacco mosaic virus in water, deoxyribonucleic acid in water and some synthetic polypeptides such as poly-γ-benzyl-L-glutamate in not too polar solvents. Stroking solutions with a flat blade until they are dry will produce films from which pieces showing high birefringence may be selected. It will probably be found that some solvents give better results than others. Polymers of high molecular weight are usually more easily oriented than are those of low molecular weight.

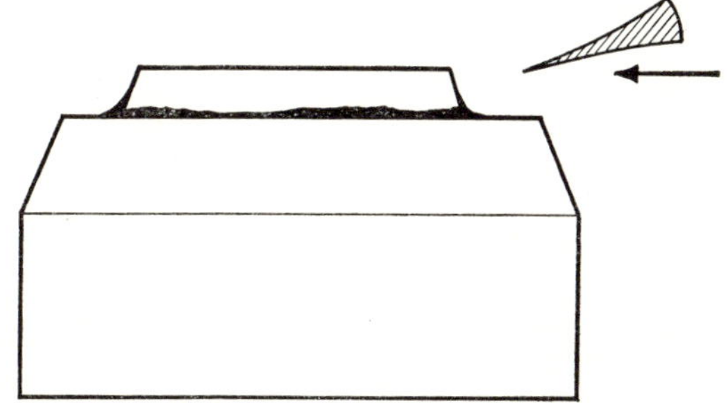

Fig. 2.8 Cutting a fibre section with a microtome knife.

SECTION CUTTING

Some polymeric materials occur in pieces much too thick for spectroscopic examination, and must be reduced in thickness without changing their physical state by solution, melting, etc. Cutting sections with a microtome is then a possibility. Softer materials may be embedded and sectioned in the usual way. With hard materials it may be more satisfactory to mount the specimen on a small block as shown in Fig. 2.8. Keratinous fibres and quills may be dealt with in this way, and the glass knife used by electron microscopists is useful. Some materials may be cemented to the block with an epoxy resin, but there is a possibility of such resins (or the hardening agents) interacting with the specimen. It was found in the author's laboratory that sections of steam-stretched horse hair failed to show the expected dichroism of β-keratin when epoxy resin was used for cementing. Monofils of some synthetic materials such as polyamides are difficult to cut in sufficiently thin sections. They may be further reduced in thickness by grinding with a fine aluminium oxide hone.

THIN FIBRES

It is possible to record the infra-red spectrum of the thin fibres used in textiles in sufficient detail for identification, but quantitative intensity measurements on thin fibres are hardly possible. In order to get a specimen of sufficient width it is necessary to make a grid of fibres, and the different parts of the radiation beam traverse very different thicknesses of polymer (part of the beam will probably not pass through any part of the fibres but will go between them). Holliday (1949) has described a jig for making a grid of fibres.

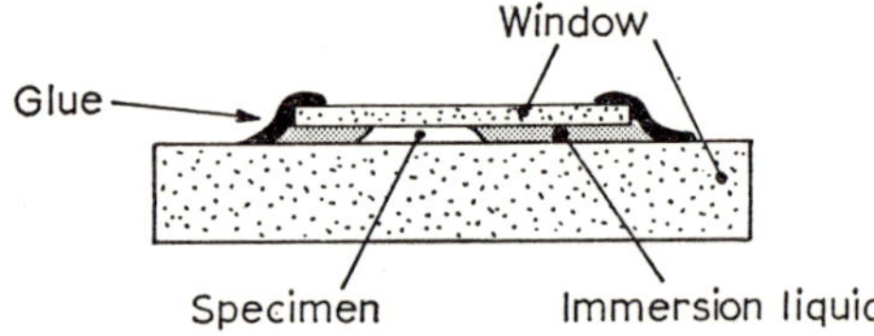

Fig. 2.9 A specimen mounted in an immersion medium to reduce reflection losses.

REDUCING SCATTER FROM THE SURFACE OF SAMPLES

A good deal of radiation may be scattered by the surface of a polymer specimen, especially if it is drawn, rolled or cut, and it is desirable to reduce this by putting the specimen in a suitable immersion material. From the visible region to 1670 cm^{-1} hexachlorobutadiene is suitable, and a fluorinated polymer of the unit—(CF_2–CFCl)—known as Fluorolube may be used from the visible region to 1350 cm^{-1}. A medicinal liquid paraffin (e.g. Nujol) (which is free from aromatic hydrocarbons) is transparent throughout the region covered by a NaCl prism except for the hydrocarbon bands 3050–2700 cm^{-1}, 1500–1340 cm^{-1} and (in thin layers) very weak absorption between 750 and 700 cm^{-1}. These materials will be found to reduce the reflection losses greatly, and not to affect most polymers. The procedure used for mounting specimens for optical microscopy may be followed. Figure 2.9 shows the arrange-

Table 2.2 Useful long-wave limit of transmission of some materials for windows (beyond which the transmission of a 1 cm thickness is less than 50%)

Fused silica	2780 cm^{-1}	Sodium chloride	590 cm^{-1}
Lithium fluoride	1610	Silver chloride	430
Calcium fluoride	1110	Potassium bromide	370
Barium fluoride	830	KRS5*	250

* Eutectic crystalline form of thallium iodide and bromide

ment. With the liquids mentioned above, a liquid protein glue such as Secotine may be used for sealing. Glues which contain phenol (recognizable by the smell) should be avoided. The useful transmission limits for materials on which specimens can be mounted are given in Table 2.2.

F DRYING AND DEUTERIUM EXCHANGE

If specimens are to be examined under specific humidity conditions, they must be enclosed in some kind of cell. It is convenient to use a cell with demountable windows, so that the specimen may be cast or mounted on the window material of the disassembled cell. (It is possible with the aid of a bent pipette to cast a film on the inside of the window of an assembled cell, but it is more difficult to control the thickness.) A cell used in the author's laboratory for such purposes is shown in Fig. 2.10. It may be made of stainless steel, but polymethylmethacrylate (Perspex, Lucite) can be used with liquids which do not attack this material. The windows, in the form of discs, are pressed against O-rings trapped in circular recesses by means of rings held by three screws and so are easily removable. The hollow base can be used for a desiccating or humidifying agent, or can be filled with heavy water to produce an atmosphere of D_2O round the specimen. This base can be shut off at will from the cell proper by means of a small polyethylene or polytetrafluorethylene plug (not shown) which has a small, tapped, blind hole so that it can be engaged by a screw thread on a rod which is inserted through the hole at the top of the cell. This allows the plug to be

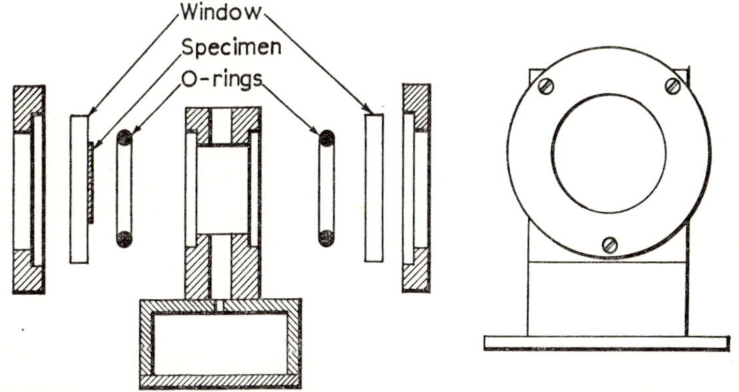

Fig. 2.10 A demountable cell in which specimens may be dried, humidified or treated with D_2O vapour. The specimen is attached to the inner face of one window.

removed and replaced. A larger plug is used to close the hole at the top.

Polymers which are penetrable by water can have the hydrogen of NH or OH groups exchanged for deuterium (see Chapter 5, section D) by submitting them to an atmosphere of D_2O from a reservoir sufficiently large to swamp the H_2O which is produced. It is necessary to keep such polymers out of contact with the atmosphere, otherwise a back-reaction will be caused by the H_2O in the atmosphere. The cell shown in Fig. 2.10 is useful for this and other purposes.

For polymers which are not easily penetrable by water (polyamides, and some polypeptides with hydrophobic side-chains) another method must be used. A solvent which normally contains OH groups, such as a carboxylic acid, is prepared in the O-deuterated form. On dissolving the polymer in such a solvent, the exchangeable groups will usually become converted to the deutero form. This operation, and the subsequent film casting must be done in the absence of H_2O vapour. It should be noted that chemical desiccants such as P_2O_5 are not satisfactory for removing H_2O in experiments on deuterium exchange, since they contain enough H_2O to contaminate the air by back-exchange (Hvidt and Linderstrøm-Lang, 1955).

G PROCEDURES

1 Wave-number calibration

Most spectrometers make use of charts on which wave-number or wavelength scales are printed. The calibration may alter slightly from time to time, either because of mechanical displacements or from changes in temperature of the prism or grating. An occasional check is quickly made by running part of the spectrum of thin polystyrene foil (about 0·025 mm). If a day-to-day variation is found, it is desirable to choose a band in this spectrum and record it superposed on the sample spectrum by inserting it in the beam at the appropriate time. The spectrum of polystyrene foil is given in Fig. 2.11.

Gases such as ammonia are often used for calibrating spectrometers. It is then necessary to use rather narrow slits, and there is something to be said for using a calibrating substance which gives absorption bands whose widths are more comparable to those of polymer bands. Such a material is indene (freshly distilled). Wave numbers in the indene spectrum have been given by Thompson (1960).

For reading off wave numbers on a strip chart, a calibrated scale on a transparent material is useful. If the spectrum to be measured

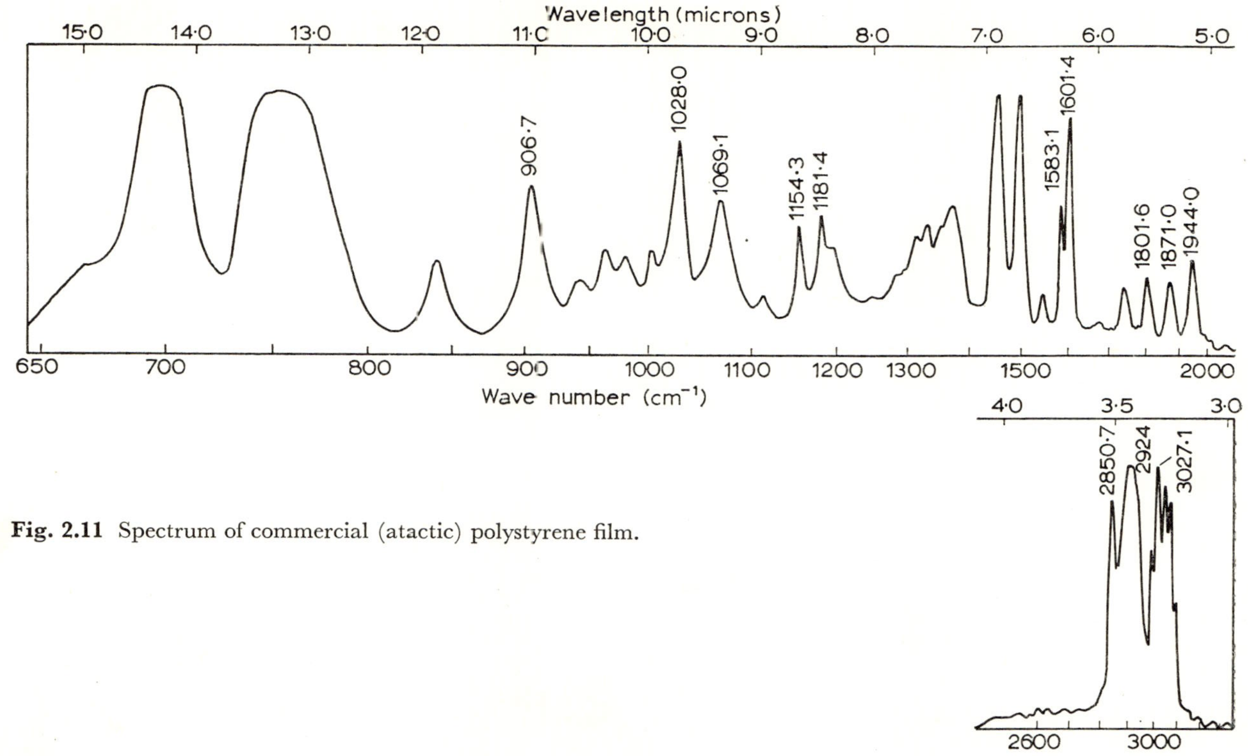

Fig. 2.11 Spectrum of commercial (atactic) polystyrene film.

contains atmospheric absorption bands, one of these should be identified and the scale set to match the wave number of this band. Alternatively, part of the polystyrene spectrum may be recorded to give a reference wave number. Fraser (1960) has given a simple method for constructing a scale. The spectrum of the calibration substance is recorded on a blank chart and a line perpendicular to the length of the chart is drawn through each peak. A distance proportional to $1/\nu^2$ is laid off along each line from a convenient base line, and the points are joined by a smooth curve, which will be almost linear. The blank scale is laid on the chart and the position of a band chosen as reference is marked. The values of ν^2 corresponding to the frequencies to be marked on the scale are located on the curve and the values of ν are inscribed directly on the scale.

2 Use of polarizers

The position and orientation of the polarizer requires consideration. Prisms introduce some polarization into an unpolarized beam, and diffraction gratings introduce a great deal at some wavelengths. Since the oriented specimens used with polarizers are birefringent, it may happen that the specimen is, in effect, put between crossed polarizers, and the apparent transmission is then to some extent dependent on the birefringence. Also, a plane polarized beam incident on a birefringent specimen will, in general, be converted into an elliptically polarized one and interpretation of the dichroic effects becomes very difficult.

These difficulties are avoided if measurements are always made with the **E** vector of the radiation either perpendicular or parallel to the slit, and also with the mean direction of the polymer molecules (the fibre axis) parallel to it. If the orientation is such that the molecules are equally distributed around the fibre axis, the polymer will be optically uniaxial. (This is true even if the crystallites within the polymer are of lower symmetry than hexagonal or trigonal, because they are so much smaller than the wavelength of the radiation.) When the **E** vector is parallel or perpendicular to the optic axis, the plane polarized wave is transmitted without change in the direction of the plane of polarization and with the arrangement described plane polarized radiation is transmitted through the whole system. Provided that I_0 as well as I are measured with the same direction of the **E** vector, the optical densities D_π and D_σ can be determined.

The polarizing properties of the grating or prism are shown by a change in the energy falling on the detector when the polarizer is rotated from the 'parallel' position (**E** vector parallel to the slit) to the 'perpendicular'. For a sodium chloride prism, the transmission for the parallel position is about 60% of that in the perpendicular position. For a grating, the relative transmission varies greatly with the setting; see, for instance, Fraser and Suzuki (1965a) whose results are given in Fig. 2.12. If the beam is not completely polarized, the proportions of parallel and perpendicular components will be altered by the dispersing element. With a silver chloride polarizer and a prism the errors can be serious if the optical density of the sample is high (Charney, 1955). For gratings, however, the effect is catastrophic. In Fig. 2.12 the transmission ratios $t_\pi/(t_\pi + t_\sigma)$ and $t_\sigma/(t_\pi + t_\sigma)$ are plotted against frequency. At 670 cm^{-1}, in the second order of the first grating, the ratio of the transmission coefficients is $0{\cdot}11/0{\cdot}89 = 1/8{\cdot}1$. Hence, with a silver chloride polarizer passing 5% of unwanted radiation (fractional polarization 0·9) the error is the same as would have been obtained with a non-polarizing spectrometer and a polarizer passing 40% of the unwanted component. The error varies greatly with frequency.

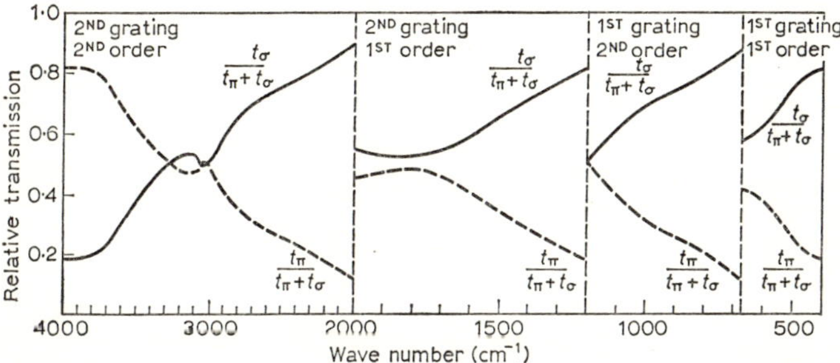

Fig. 2.12 Transmission coefficients of the Beckman IR9 spectrometer for polarized radiation. (Fraser and Suzuki, 1965a)

Charney (1955) has suggested orienting the **E** vector at 45° to the slit, so making the transmission of the spectrometer the same for the two positions of the polarizer. This requires the specimen to be large enough to place it also with its fibre axis at 45° to the slit. With a selenium polarizer, the efficiency is so high that the error caused by the partial polarization is very much less. Fraser and Suzuki have used such a polarizer with a Beckmann IR9 spectrometer, placing it in the recombined sample and reference beams as shown in Fig. 2.1. With this arrangement the measurements are straightforward and the fibre axis of the specimen is made parallel to the slit.

3 Representation of results

If the record produced by the spectrometer is on a chart with a wave-number scale, it will be suitable for most purposes. For accurate measurements of dichroic ratios it may be necessary to redraw the spectrum, unless the recorder is of the kind which gives a deflection proportion to the optical density. On such a record, allowance for background absorption can be made, and some allowance for imperfectly resolved bands may be attempted. The conversion of a spectrum recorded as transmittance to optical density may be done by calculation of log I_0/I, reading I_0 and I from the curve, but it is much more quickly done with a scale as shown in Fig. 2.13. Vertical lines (constant wave number) may first be drawn

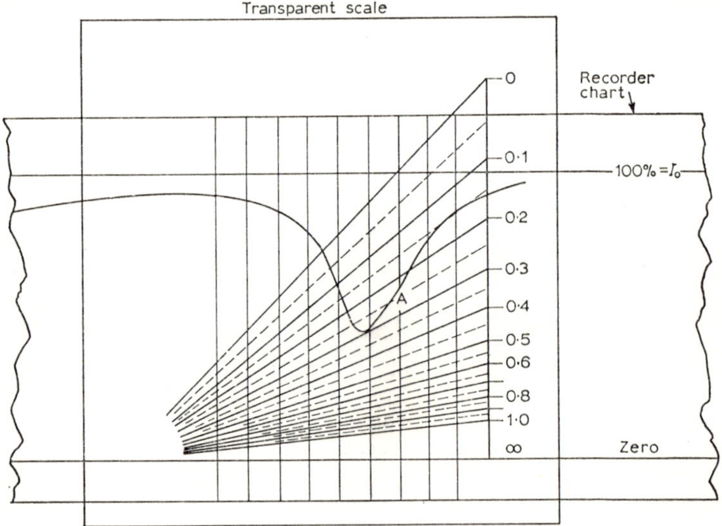

Fig. 2.13 A transparent scale for the direct reading of optical densities from spectral transmission records. The optical density at A is 0·23.

on the record at suitable intervals, and the lines for zero and 100% transmission should also be drawn. The 45° line of the scale is set on the 100% transmission line at a chosen wave number, with the horizontal line (density infinite) on the zero transmission line at the same wave number. The optical density can then be read off. The process is greatly speeded up if the scale is constrained so that it has a translational movement only, without rotation (as in a draughting machine on a drawing board). The spectrum is redrawn with optical density as a function of wave number. The apparent optical density in a region well away from absorption bands should be subtracted before the results are plotted. When using polarized radiation the apparent optical density must be measured for both the parallel and the perpendicular positions of the polarizer. The apparent optical density represents losses from irrelevant factors such as reflection and scattering, or beam obscuration. Many of the spectra in this book have been plotted in the way described.

Errors

Instrumental factors such as instability of zero, random fluctuations of the indicator ('noise'), lack of proportionality between actual transmission and the measured value are outside the scope of this book. There remain, however, some sources of error which may be appropriately discussed.

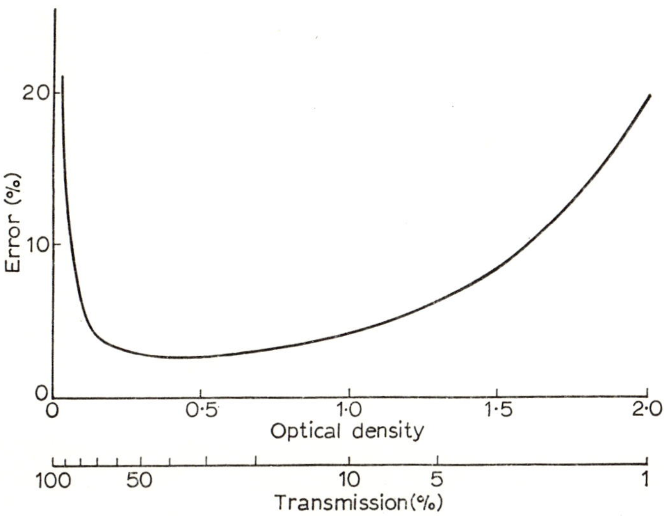

Fig. 2.14 The percentage error in optical density caused by a constant error (such as a base-line error) in measuring the transmitted intensity. Error assumed = 1% of I_0. After Robinson (1951)

THICKNESS OF SPECIMEN

A given percentage error in the transmission measurement will produce a percentage error in the measured optical density which varies greatly with that density. Figure 2.14 (Robinson, 1951) shows that at high and at low transmissions the percentage error in the optical density rises sharply. It is desirable to choose the specimen thickness so that the transmission lies in the range 20–60%.

NON-UNIFORMITY OF THICKNESS

Probably the greatest errors in the spectroscopy of polymers are caused by non-uniformity of thickness. Since it is often difficult to make uniform specimens, these errors are of frequent occurrence. The effect may often be seen in the stronger bands, whose peaks are flattened if the thickness is not uniform. The extreme is met when there are holes in an otherwise uniform specimen. Jones (1952) has considered the error in measured optical density when there is a fraction X of uniformly absorbing material, with a fraction $(1 - X)$ of clear space; the relation between measured and true optical density is shown in Fig. 2.15. The percentage error increases rapidly with increasing true optical density.

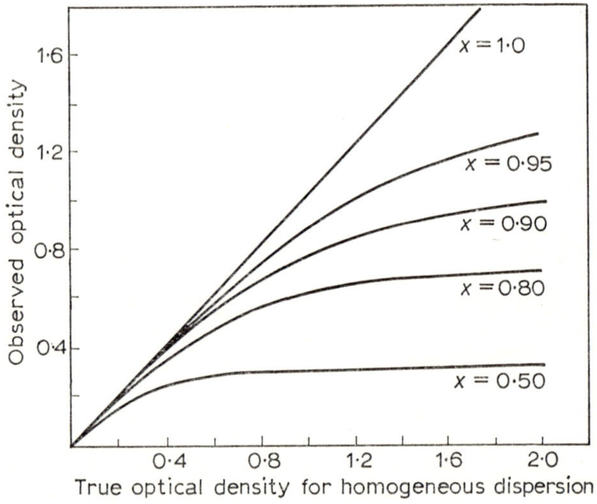

Fig. 2.15 Errors caused by specimen non-uniformity. True optical density of specimen plotted against observed optical density for different values of X, where $(1 - X)$ is the fraction of the incident radiation passing through clear spaces between the absorbing material. (Jones, 1952)

IMPERFECTLY RESOLVED BANDS

Because the absorption bands of polymers are often broad, it is common for one band to overlap its neighbours, and a weak band will often appear as a 'shoulder' on a stronger neighbouring band. If the spectrum is drawn on an optical density scale, a graphical resolution can be attempted, the object being to produce two bands of approximately symmetrical form whose ordinates, when added, give the observed unresolved spectrum. Figure 2.16 shows an unresolved band whose components have been separated in this way. The frequency of the weaker peak is more reliably measured after such a separation has been carried out. Estimates of dichroic ratios

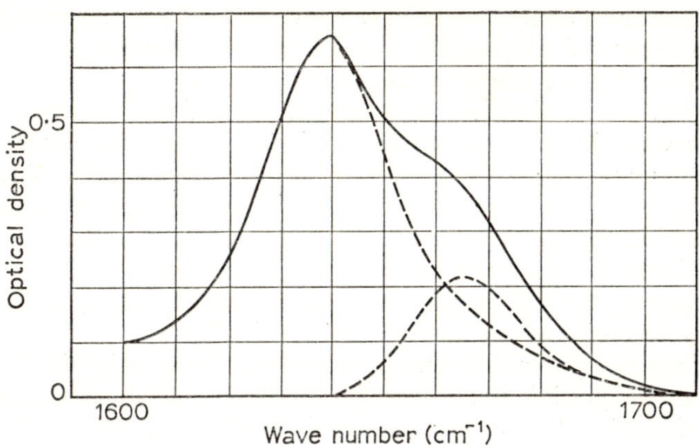

Fig. 2.16 Graphical resolution. The broken curves show the component bands, whose ordinates add to give the full curve.

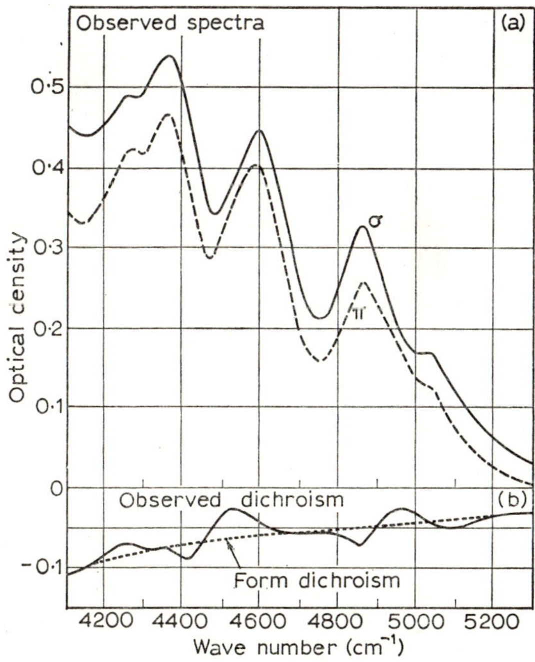

Fig. 2.17 (**a**) An example of form dichroism observed with a bundle of wool fibres immersed in hexachlorobutadiene. (**b**) Observed dichroism (full curve) and assessment of form dichroism (dotted curve). (Fraser, 1960)

on unresolved bands may be in error even after the components have been separated graphically, but are not so meaningless as those based on the ordinates of unresolved peaks.

Systematic methods for separating unresolved bands by the use of a computer have been described by Fraser and Suzuki (1966) and by Pitha and Jones (1966). Suitable functions for representing the shapes of the bands are chosen, and the parameters are determined for which the mean square deviation between the observed and calculated spectra is a minimum. It will be realized that for such procedures, the instrumental resolution should be good, so that an accurate profile of the band complex is obtained.

When the spectrum has many unresolved bands as in Fig. 2.17, it becomes impossible to apply such procedures, and the height of the individual bands cannot be measured. If polarized radiation is used with the object of making measurements of infra-red dichroism, it is likely that the reflection and scattering losses will depend on the direction of the electric vector of the radiation. This is particularly noticeable if the specimen has a fibrillar structure, and is a consequence both of the inherent birefringence of oriented materials and also of the so-called form birefringence of fibrous specimens. Fraser (1960) has suggested a procedure for making some allowance for such effects. The difference in the optical density of the two curves is plotted, and a smooth curve drawn through the points

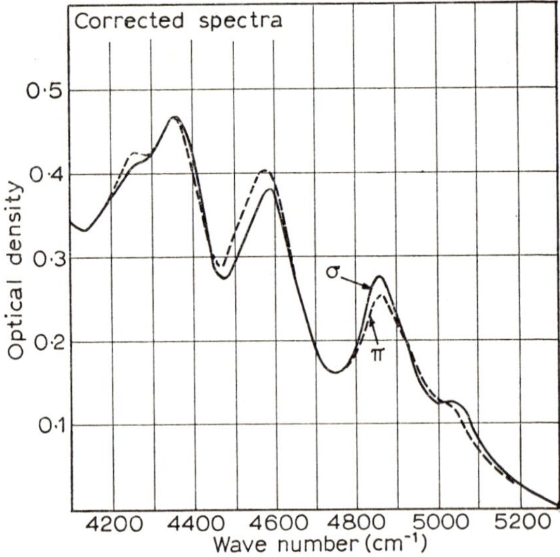

Fig. 2.18 Spectrum of Fig. 2.17 after subtraction of dotted curve from perpendicular spectrum. (Fraser, 1960)

as shown (Fig. 2.17). The correction curve is then subtracted from the upper curve and the spectrum shown in Fig. 2.18 results. Although accurate measurements of the dichroic ratio D_π/D_σ cannot be read off, the dichroic character of the several peaks is shown much more clearly.

CHAPTER 3

Vibrations in Chain Molecules

A ISOLATED POLYMER CHAINS

1 Symmetry types

A molecule containing N atoms has $3N$ degrees of freedom, requiring $3N$ independent coordinates to specify the positions of its atoms. The number of normal modes may be shown to be $3N$, of which in general six are of zero frequency, and correspond to translations and rotations of the system as a whole (see, for instance, Herzberg, 1945). The spectrum of a polymer containing a large number of atoms might therefore be expected to be very complicated. Although the number of vibration frequencies of a randomly coiled polymer chain is very large, many of them are so nearly identical that the corresponding absorption bands are not resolved, so the spectrum appears much more simple than it really is. If the polymer chain is regularly arranged in a series of repeating units (in the crystallographic sense) of identical conformation, then at least for an infinite chain a real simplicity results. It may be shown that in this case only those vibrational modes can be infra-red or Raman active for which all such units are vibrating in phase. These are the vibrations proper to the assembly of atoms in the repeating unit. If there are r atoms in this unit, the number of vibrations (including those of zero frequency) will be $3r$. Translations of the whole chain account for the three modes of zero frequency, and rotation around the chain axis is a zero frequency mode in which all units have identical motion. But rotations around axes perpendicular to the chain axis do not allow identical movements within all units unless the links between one unit and the next are extended, and then the motion will not be of zero frequency. We conclude that for an infinite isolated chain the number of potentially active modes of non-zero frequency is $3r - 4$. If the chain is not of infinite length, then all the $3N$ normal modes are potentially active but, as will be seen later, most of these are of very low intensity in the polymers of high molecular weight which are effectively of infinite length.

When the system is executing a single normal vibration, all the atoms move with the same frequency, and in non-degenerate modes it may be shown that the ratio of the displacements of the various atoms remains constant during the vibration. If a symmetry operation is carried out, this ratio (which depends on masses and on force constants) is unaltered. Hence for non-degenerate modes a symmetry operation must either change the sign of none of the displacements or of all of them (in degenerate modes other changes are possible). In the first case, the pattern of atomic positions remains unchanged and the mode is *symmetrical* for the particular operation which has been carried out; in the second case, the phase of the motion is altered by π radians and the mode is *antisymmetrical* for the operation.

The normal modes may be grouped into symmetry classes or species according to their behaviour when the various symmetry operations are applied. Not all combinations of symmetry and antisymmetry are possible because these operations are not independent, the presence of some elements implying the presence of others. For a discussion of this subject, including the consideration of degenerate vibrations, the reader is referred to works on the infra-red spectra of simple molecules such as that of Herzberg (1945). For simple molecules the symmetry elements form a *point* group (all symmetry planes, axes and centres pass through one point). The symmetry species of the normal vibrations for the point groups are given in tables similar to Table 3.1, which is for the combination of symmetry properties designated by the point group symbol $D_{2h} \equiv V_R$. The following symmetry elements are present in this group:

E the identity (the molecule is not operated on)
$C_2(x)$ a 2-fold rotation (cyclic) axis along x
$C_2(y)$ a 2-fold rotation axis along y
$C_2(z)$ a 2-fold rotation axis along z
i a centre of symmetry or inversion element
$\sigma(xy)$ a mirror plane in the xy plane
$\sigma(yz)$ a mirror plane in the yz plane
$\sigma(xz)$ a mirror plane in the xz plane

In Table 3.1 the entries $+1$ and -1 (called the characters) indicate that the vibrations are symmetric or antisymmetric, respectively, to the symmetry operation at the head of the column. The letter A is used for species which are symmetric, B for species antisymmetric, to a rotation axis. The various species of types A and B are distinguished from each other by number subscripts, B_1, B_2, etc. Subscripts g and u (gerade and ungerade—even and odd) are used to indicate symmetry and antisymmetry, respectively, to inversion at a centre of symmetry. Table 3.1 also shows to which of the symmetry species the zero frequency vibrations belong, translations and rotations being indicated by T and R with the corresponding axis as subscript. The use of this information is explained later.

Table 3.1 Symmetry species for the point group D_{2h} and for the isomorphous line group of the polyethylene chain

Point group D_{2h}	E	$C_2(y)$	$C_2(x)$	$C_2(z)$	i	$\sigma(xz)$	$\sigma(yz)$	$\sigma(xy)$		
Line group	E	$C_2(y)$	$C_2(x)$	$C_2^s(z)$	i	$\sigma(xz)$	$\sigma_g(yz)$	$\sigma(xy)$	zero* modes	n†
A_g	+1	+1	+1	+1	+1	+1	+1	+1		3
A_u	+1	+1	+1	+1	−1	−1	−1	−1		1
B_{1g}	+1	−1	−1	+1	+1	−1	−1	+1	R_z	3
B_{1u}	+1	−1	−1	+1	−1	+1	+1	−1	T_z	2
B_{2g}	+1	+1	−1	−1	+1	+1	−1	−1		2
B_{2u}	+1	+1	−1	−1	−1	−1	+1	+1	T_y	3
B_{3g}	+1	−1	+1	−1	+1	−1	+1	−1		1
B_{3u}	+1	−1	+1	−1	−1	+1	−1	+1	T_x	3

* This column gives the species of the zero frequency modes.
† Total number of vibration modes for an isolated, extended, infinite polyethylene chain (including zero modes). Five are infra-red active.

Table 3.1 has been included because it (and similar tables for other points groups) can be used to classify the vibrations in linear polymers. Stereo-regular polymers in a regular chain conformation of infinite extent (that is, having a one-dimensional repeat of pattern) have symmetry elements not present in point groups. These are: the unit translation, screw rotation axes and glide planes, in addition to an infinite number of rotation axes and mirror planes if these are present in the repeating units. The symmetry operation of a p-fold screw axis is a rotation of $360°/p$ followed by a translation along this axis. The length of the unit translation (axial repeat) is p times the translation associated with the screw axis. The symbol for the screw axis is $C_p^s(x)$, the letter in parentheses indicating the coordinate axis whose direction coincides with the screw axis. The symmetry operation of a glide plane is reflection at the plane followed by translation in a direction which is parallel to the plane. The symbol is $\sigma_g(xy)$. These elements form *line* groups (all axes, centres and planes of symmetry pass through a line, the chain axis). As mentioned earlier, the potentially infra-red and Raman active modes depend only on the structure of the unit of repeat. The symmetry species can be predicted from considerations of the symmetry elements of this unit. These elements, together with the unit translation (equivalent to the identity E) form a sub-group of the line group, sometimes called a unit cell group or a factor group. Since it may include screw axes and glide planes this unit cell group is not a point group. There is, however, a point group which is

Table 3.2 Multiplication table for the point group D_{2h}

	E	$C_2(y)$	$C_2(x)$	$C_2(z)$	i	$\sigma(xz)$	$\sigma(yz)$	$\sigma(xy)$
E	E	$C_2(y)$	$C_2(x)$	$C_2(z)$	i	$\sigma(xz)$	$\sigma(yz)$	$\sigma(xy)$
$C_2(y)$	$C_2(y)$	E	$C_2(z)$	$C_2(x)$	$\sigma(xz)$	i	$\sigma(xy)$	$\sigma(yz)$
$C_2(x)$	$C_2(x)$	$C_2(z)$	E	$C_2(y)$	$\sigma(yz)$	$\sigma(xy)$	i	$\sigma(xz)$
$C_2(z)$	$C_2(z)$	$C_2(x)$	$C_2(y)$	E	$\sigma(xy)$	$\sigma(yz)$	$\sigma(xz)$	i
i	i	$\sigma(xz)$	$\sigma(yz)$	$\sigma(xy)$	E	$C_2(y)$	$C_2(x)$	$C_2(z)$
$\sigma(xz)$	$\sigma(xz)$	i	$\sigma(xy)$	$\sigma(yz)$	$C_2(y)$	E	$C_2(z)$	$C_2(x)$
$\sigma(yz)$	$\sigma(yz)$	$\sigma(xy)$	i	$\sigma(xz)$	$C_2(x)$	$C_2(z)$	E	$C_2(y)$
$\sigma(xy)$	$\sigma(xy)$	$\sigma(yz)$	$\sigma(xz)$	i	$C_2(z)$	$C_2(x)$	$C_2(y)$	E

isomorphous with the unit cell group. This may be found by writing out the multiplication table for the sub-group and comparing it with the multiplication tables of the thirty-two crystallographic point groups. One of these will be found to have a one-to-one correspondence with the multiplication table for the unit cell group. This is made clearer by Tables 3.2 and 3.3 which give the multiplication tables for the point group D_2h and the isomorphous unit cell group for linear polyethylene. The tables are made by writing down all the symmetry elements (including the identity E) in double entry, and entering the symmetry operation which results from applying first the operation at the head of the column, then the operation given at the end of corresponding row. Table 3.3 includes the 2-fold screw axis about the z axis (symbol $C_2{}^s(z)$) and the glide plane parallel to yz (symbol $\sigma_g(yz)$). Except that these symbols are replaced in Table 3.2 by $C_2(z)$ and $\sigma(yz)$, the tables are identical and the groups are isomorphous. The symmetry species for the normal vibrations of linear polyethylene may be read from Table 3.1. A detailed discussion of this procedure, with examples, is given by Zbinden (1964) and by Krimm (1960).

Table 3.3 Multiplication table for the line group of the polyethylene chain (isomorphous with D_{2h})

	E	$C_2(y)$	$C_2(x)$	$C_2{}^s(z)$	i	$\sigma(xz)$	$\sigma_g(yz)$	$\sigma(xy)$
E	E	$C_2(y)$	$C_2(x)$	$C_2{}^s(z)$	i	$\sigma(xz)$	$\sigma_g(yz)$	$\sigma(xy)$
$C_2(y)$	$C_2(y)$	E	$C_2{}^s(z)$	$C_2(x)$	$\sigma(xz)$	i	$\sigma(xy)$	$\sigma_g(yz)$
$C_2(x)$	$C_2(x)$	$C_2{}^s(z)$	E	$C_2(y)$	$\sigma_g(yz)$	$\sigma(xy)$	i	$\sigma(xz)$
$C_2{}^s(z)$	$C_2{}^s(z)$	$C_2(x)$	$C_2(y)$	E	$\sigma(xy)$	$\sigma_g(yz)$	$\sigma(xz)$	i
i	i	$\sigma(xz)$	$\sigma_g(yz)$	$\sigma(xy)$	E	$C_2(y)$	$C_2(x)$	$C_2{}^s(z)$
$\sigma(xz)$	$\sigma(xz)$	i	$\sigma(xy)$	$\sigma_g(yz)$	$C_2(y)$	E	$C_2{}^s(z)$	$C_2(x)$
$\sigma_g(yz)$	$\sigma_g(yz)$	$\sigma(xy)$	i	$\sigma(xz)$	$C_2(x)$	$C_2{}^s(z)$	E	$C_2(y)$
$\sigma(xy)$	$\sigma(xy)$	$\sigma_g(yz)$	$\sigma(xz)$	i	$C_2{}^s(z)$	$C_2(x)$	$C_2(y)$	E

2 Number of vibrations of each symmetry type

In a molecule with symmetry elements, the atoms can be arranged in sets of equivalent atoms. Atoms are 'equivalent' if a symmetry operation will carry a (non-vibrating) atom over into the position occupied by another non-vibrating atom. For example, there are two equivalent carbon atoms and four equivalent hydrogen atoms in the extended polyethylene chain (Fig. 3.1). The equivalent atoms in

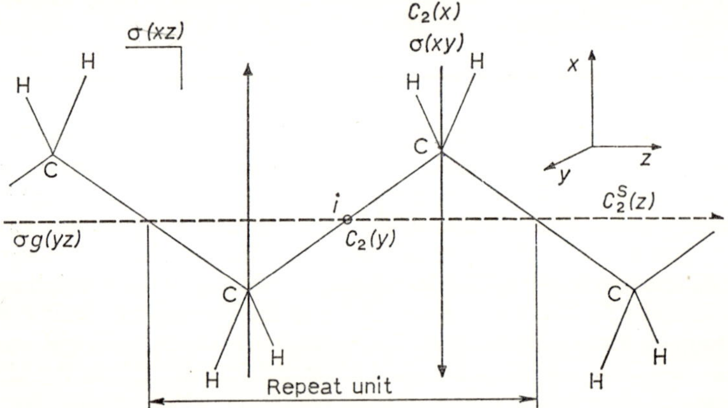

Fig. 3.1 Symmetry elements in an extended polyethylene chain.

a set can all be produced from a single atom by successive symmetry operations applied to the equilibrium (non-vibrating) state. The number of equivalent nuclei in a set is a maximum if the representative atom is in a general position (not on any symmetry element), and is less than the maximum if this atom is on one or more of the symmetry elements. If the atom lies on all the symmetry elements the set contains only one atom.

When the system is vibrating in a single non-degenerate normal mode, a symmetry operation will or will not leave the system unchanged according as the mode is symmetric or antisymmetric for that operation. For a given symmetry type (species) the displacements of all the atoms in a set are fixed by the displacements of one of them, hence the atoms in a set can at most contribute three degrees of freedom to each non-degenerate species. When the representative atom lies on one or more symmetry elements it may have less than three degrees of freedom. For example, if an atom lies on a plane of symmetry, the motion can be symmetric with respect to this plane only if it is in the plane, and then cannot contribute more than two degrees of freedom. If the motion is to be antisymmetric with respect to the plane, it must be normal to the plane and can contribute only one degree of freedom. If the representative atom lies on a 2-fold rotation axis, motion along this axis will be symmetric with respect to it. If the motion is perpendicular to a 2-fold axis, it is antisymmetric with respect to the axis.

The symmetry elements of an infinite extended polyethylene chain are shown in Fig. 3.1 and the symmetry species are given in Table 3.1. There are two sets of equivalent atoms. The two carbon atoms lie on the symmetry elements $C_2(x)$, $\sigma(xy)$ and $\sigma(xz)$. The four hydrogen atoms lie on $\sigma(xy)$. The normal modes are shown in Fig. 3.2.

Vibrations of species A_g must be symmetrical for all operations, (totally symmetric). The hydrogen atoms, being on the $\sigma(xy)$ plane must move in this plane during an A_g vibration and contribute two degrees of freedom in consequence. The carbon atoms can move symmetrically with respect to the $C_2(x)$ operation only if they move along this axis, and so have one degree of freedom. There are therefore $2 + 1 = 3$ vibrations of species A_g. Neither the translations T_x, T_y, T_z nor rotation around the z axis (R_z) is compatible with all the symmetry operations. The three vibrations are therefore genuine (non-zero) vibrations.

For species A_u the motion is antisymmetric for the mirror and glide planes and for the centre of symmetry i, and symmetric for the rotation axes. The hydrogen atoms must therefore move normal to $\sigma(xy)$ and hence have only one degree of freedom. It is easily seen that if the atoms on one CH_2 group move simultaneously along the z axis in opposite directions the other symmetry requirements are met. The carbon atoms must move along $C_2(x)$ if the motion is to be symmetric with respect to this axis, but antisymmetry with respect to $\sigma(xy)$ requires motion normal to this plane and the two requirements cannot be met with a finite motion. Hence in the species A_u the carbon atoms do not move (zero degree of freedom) and there is therefore only one mode and it is of non-zero type.

In species B_{2g} the hydrogen atoms move perpendicular to the $\sigma(xy)$ plane, with respect to which the motion is antisymmetrical. The two atoms in the CH_2 group move simultaneously along the z axis, giving a motion which is antisymmetrical to the $C_2(x)$ axis as required. The other requirements can be met by giving the other CH_2 group the appropriate motion and the hydrogen atoms contribute one degree of freedom. The carbon atoms must move perpendicular to $\sigma(xy)$ and this motion is antisymmetric with respect to $\sigma(xy)$ and $C_2(x)$. They contribute one degree and so there are two vibrations of species B_{2g}, both genuine. In similar fashion the number of vibrations in the eight symmetry types can be found. They include the zero vibrations T_x, T_y, T_z and R_z shown in Fig. 3.2 (see also Table 3.1). More formal methods for determining the

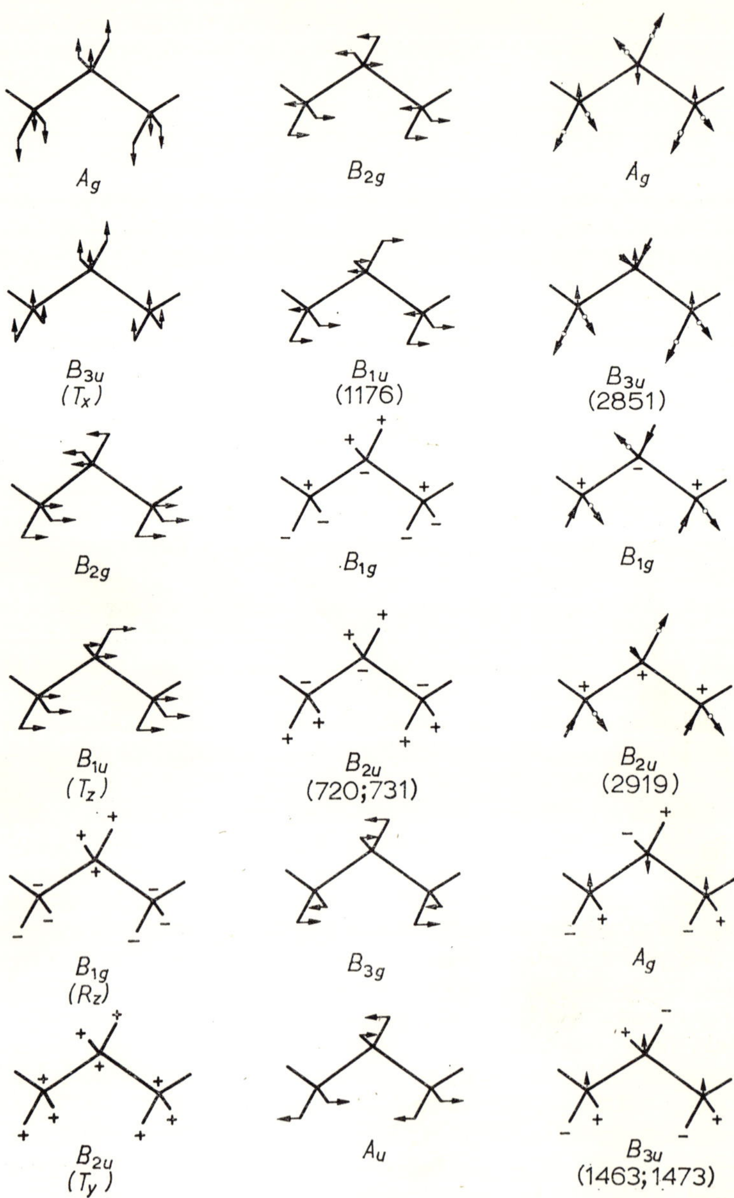

Fig. 3.2 Symmetry modes of a single polyethylene chain. (Krimm, Liang and Sutherland, 1956; the designation has been altered to conform to the symbols in Table 3.4)

distribution of vibrational modes over the symmetry types (including degenerate modes) are given by Herzberg (1945) and Zbinden (1964).

3 Phase relations within a repeating unit of a chain

Polymer chains in a regular conformation usually have several identical groups (each containing a number of atoms) within the repeating unit of the chain, and certain phase relations exist between the motions of the corresponding atoms within the groups. This is included in the symmetry analysis above, but it is sometimes helpful to consider these relations explicitly. If there are q groups related by a q-fold screw axis in the repeating length, vibrational modes occur in which the phase angle ϕ relating the motion within one group to the motion in the neighbouring group is given by

$$\phi = \pm\frac{2\pi r}{q} \quad \text{where} \quad r = 0, 1, 2, \cdots, q - 1.$$

In these vibrations, corresponding motions in all the repeating units of the chain are in phase, so the vibrational modes are potentially infra-red and Raman active. In one mode all the groups move in phase, in another neighbouring groups are $2\pi/q$ out of phase and so on. When $q = 2$ the vibrations are either in phase or differ by π and so can be described as symmetrical or antisymmetrical with respect to the 2-fold screw axis. If Fig. 3.2 is examined it will be seen that in half of the eighteen modes the two CH_2 groups have motions which are symmetrical to the 2-fold screw axis $C_2^s(z)$ and so are in phase, and in the other half the motions are antisymmetrical (phase difference π). When $q > 2$ the vibrations are degenerate because the phase difference between the motions of neighbouring groups may be either positive or negative. This gives two modes of identical frequency for each value of r above. Only when r is zero (all groups moving in phase) is the motion non-degenerate. For this species, the infra-red active mode is one in which the transition moment is along the helix axis. For the degenerate modes which are infra-red active, $r = 1$ (the phase difference between vibrations in neighbouring groups is then equal to the angular rotation which carries one group into the next) and the transition moment is normal to the helix axis (Higgs, 1953; Liang and Krimm, 1956).

A fruitful application of the concept of phase relations between vibrations within a repeating unit has been made by Miyazawa (1960) and Miyazawa and Blout (1961) and this will be considered in Chapter 5. Here it need only be said that Miyazawa and Blout have satisfactorily explained how the frequency of the Amide I band in polypeptides and proteins depends on chain conformation, by invoking interactions between neighbouring peptide units.

4 Effect of finite chain length

When a polymer chain is of finite length, many more potentially active vibrations occur than those belonging to the repeating unit. A general account of this problem is given by Zbinden (1964) in terms of the motions of a linear array of N coupled dipoles of frequency ω_0. The number of modes of the system is N, but not all are active. The four models considered are parallel and antiparallel arrangements of dipoles, each with two free or with two fixed ends. The mode giving the most intense absorption band has a frequency ω_0 for parallel and near ω_0 for antiparallel arrangements. The other modes are of higher frequency. For parallel arrangements with free ends, only one frequency ω_0 is active. In the other arrangements the intensity of absorption from the modes of higher frequency is low. As N increases, the component modes become more closely spaced, and coalesce when N becomes infinite.

In a finite polymer chain (say of polyethylene) every individual CH_2 group is considered as an oscillator with a dipole moment changing in the direction of the transition moment for one of the internal vibrational modes of the group. For the other internal modes the group is replaced by the corresponding oscillator. For example, the rocking vibration in CH_2 has a frequency *c.* 720 cm^{-1} and the dipole moment change is perpendicular to the chain axis (see Fig. 3.2). The progression of bands seen in Fig. 3.3, starting

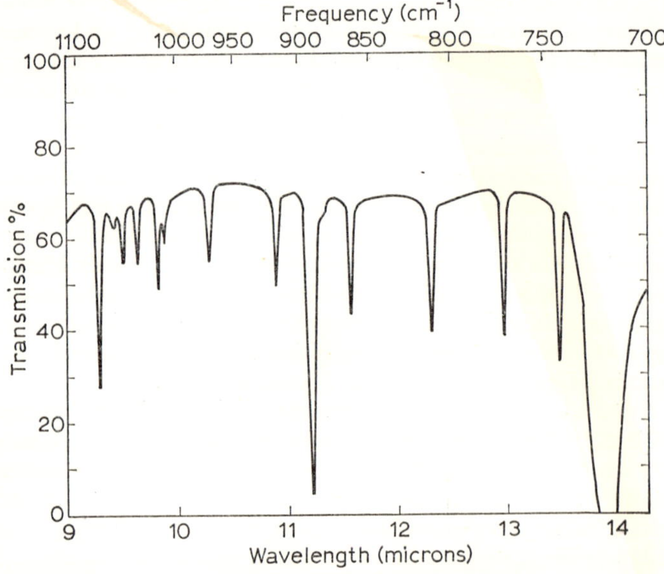

Fig. 3.3 Spectrum of the *n*-hydrocarbon $C_{24}H_{50}$ at -160°C. (Snyder, 1957)

with the very strong band at 720 cm^{-1} and with much weaker bands at intervals of about 40 cm^{-1} (not quite regularly spaced) is an example of the modes of a finite chain. In this case there are twenty-two CH_2 groups which appear to vibrate with fixed ends (if the ends were free only the strongest band would appear). It is found that the frequency of the strongest band due to the rocking mode is almost constant in the hydrocarbon series when the number of CH_2 groups is greater than six. The intervals between the bands of the progression depend markedly on the length of the chain, however. Hence it is to be expected that in a polymer with a range of chain lengths, only the strongest band will be recognized (since all chains will contribute to it, whereas the weaker bands will cover a wide spectrum of frequencies). It may be expected that some asymmetry will be given to absorption bands of polymers of sufficiently low molecular weight, the bands being broadened on the high-frequency side, but the effect is not noticeable in the spectra of high polymers.

The vibrational spectra of the crystalline *n*-paraffins, which illustrate the effect of finite chain length, have been studied by Snyder (1960).

5 Intensities, selection rules and polarization properties

Infra-red absorption bands are associated with vibrations during which the electric dipole moment changes. There need be no permanent dipole moment, and indeed very strong absorption bands can occur in substances with no permanent dipole moment (such as carbon tetrachloride, for instance). Apart from the properties of the vibration, to be considered below, the local separation of the electric charges on the atoms forming a bond (or the polar character of that bond) will be a cause of strong absorption when these bonds are deformed or stretched. So the absorptions associated with the bond stretching of such groups as CCl, C=O, OH, NH and of ionized groups will be strong, whereas those of CH, CC and SH will be weak.

In classical theory, the intensity of absorption depends on the square of the rate at which the dipole moment changes with the coordinates. If the motion is simple harmonic, only one frequency of absorption occurs for each normal mode. In fact, the motion is never exactly simple harmonic, and so contains not only the fundamental vibration ν_i of a particular normal mode i, but also overtones $2\nu_i$, $3\nu_i$, etc., and in addition combinations with other normal modes $\nu_i + \nu_k$, $\nu_i - \nu_k$, $2\nu_i + \nu_k$, etc. The spectrum therefore contains overtone and combination bands as well as the fundamentals. Because the anharmonicity is slight, the rate of change in dipole moment with the coordinates is small for these higher frequencies, which are correspondingly weak.

The Raman effect is observed when a molecule is irradiated with light of frequency ν and part is re-radiated with a change of frequency $\delta\nu$. The frequency change is related to the vibrational frequencies of the molecule. A vibrational mode of a molecule will be Raman active if a change of the electric polarization occurs during the vibration. The Raman effect has not been greatly used in the study of polymers, largely because of experimental difficulties, though it has furnished important information about the vibrations in polyethylene and a few other polymers. Its importance lies in the fact that vibrations which are not active in the infra-red are often active in the Raman effect. It is not considered in detail here. An account of the observation of the Raman effect in polymers has been given by Nielsen (1964) and more recently by Matsui, Kubota, Tadokoro and Yoshihara (1965).

In the quantum theoretical treatment absorption bands occur with the transition of the molecule from one quantum state to another. If v', v'' stand for all the vibrational quantum numbers in the upper and lower states, respectively, then for fundamentals v'' is zero for all modes, and the ν_ith fundamental is excited when the corresponding quantum number v_i changes from 0 to 1. Overtones of ν_i occur when v_i changes from 0 to 2, 3, etc. Combination bands result from a simultaneous change in the quantum number of two or more normal modes.

The probability of a vibrational transition $v' \longleftrightarrow v''$, on which the intensity of a spectral band depends, is proportional to the square of the transition moment $[\mathbf{M}]^{v'v''}$

$$[\mathbf{M}]^{v'v''} = \int \psi_v' \psi_v''^* \mathbf{M}\, d\tau.$$

$\mathbf{M}$ is the dipole moment vector and has components M_x, M_y and M_z; ψ_v' and ψ_v'' are the vibrational eigenfunctions of the vibrational levels and the integral extends over the whole of space. The asterisk denotes the complex conjugate quantities. Provided that the coordinates describing the atomic motion are chosen in a particular way (normal coordinates), the vibrational eigenfunction for the molecule in any allowed vibrational state can be represented in approximation by the product of a number of eigenfunctions, one for each normal mode. Then $\psi = \psi_1(\xi_1) \cdot \psi_2(\xi_2) \ldots \psi_{3N}(\xi_{3N})$. Here the suffixes stand for the 1st, 2nd ... $3N$th normal modes and the ξ's are the corresponding normal coordinates (Herzberg, 1945). In the harmonic oscillator approximation, the vibrational quantum number v_i for the ith vibrational mode can change during a transition involving absorption or emission of radiation by ± 1. For these allowed transitions, the ψ's for all the other modes remain unaltered and may be treated as constant factors. If the vibrations are anharmonic, the selection rule $\Delta v = \pm 1$ does not hold.

If the molecule has no symmetry elements there is no other restriction for transitions (though the intensity of absorption accompanying some transitions may be low). For molecules with symmetry, however, the above integral will be zero for some transitions.

For a finite probability of transition (that is, for the transition to be 'allowed'), at least one of the components

$$\psi_{v}'\psi_{v}''^{*}M_{x}, \quad \psi_{v}'\psi_{v}''^{*}M_{y}, \quad \psi_{v}'\psi_{v}''^{*}M_{z}$$

must be totally symmetrical (must not change for any of the symmetry operations which can be carried out on the non-vibrating system). It is obvious that if this product were to be antisymmetrical with respect to any symmetry operation the corresponding integral would be zero. This selection rule holds alike for harmonic and anharmonic oscillations. If $\psi_{v}'\psi_{v}''^{*}$ has the same symmetry properties (that is, if it belongs to the same symmetry type or species) as one of the components of **M**, their product will be totally symmetric for that component. This is easily seen for non-degenerate modes, where the result of a symmetry operation is either no change or just a change of sign, but it may also be shown that for degenerate modes the product of two species is totally symmetric or contains a totally symmetric component only when the two species are the same.

During fundamental transitions the quantum number of one vibrational mode changes from 0 to 1. The function ψ_{v} is an odd function of the coordinate for odd values of v and changes sign when the sign of the coordinates is reversed. For a vibration which is antisymmetrical for a certain symmetry element, therefore, the function will change sign on carrying out the corresponding symmetry operation and so be antisymmetrical for odd values of v. For even values of v, ψ does not alter on changing the sign of the coordinates. Hence for an antisymmetrical vibration the eigenfunctions for the 0, 1, 2, 3, etc., vibrational levels will be alternately symmetrical and antisymmetrical, the sequence being represented by $+1$, -1, $+1$, -1, etc. For a symmetrical vibration the eigenfunction is of course symmetrical, and so will be totally symmetrical for the lower level of a fundamental transition (in which v'' is zero). For such a transition, therefore, the product $\psi_{v}'\psi_{v}''$ will belong to the same symmetry species as the vibration associated with the transition $v' \rightleftarrows v''$. Now the components of **M** along the x, y and z axes behave in the same way with respect to symmetry operations and hence belong to the same symmetry species as the zero vibrations T_{x}, T_{y} and T_{z}. The symmetry species which contain one of these zero vibrations are the only ones which are infra-red active as fundamentals, for only in those species can the product $\psi_{v}'\psi_{v}''^{*}\mathbf{M}$ have a component which is totally symmetrical. Thus T_{x}, T_{y} and T_{z} give the directions of the transition moment for fundamental transitions.

To illustrate these relations the fundamental transitions in an isolated, extended chain of polyethylene will be considered.

6 Spectrum of polyethylene considered as an extended isolated chain

Table 3.1 gives the symmetry elements and vibrational species for linear polyethylene. Applying the criteria for infra-red activity, it is seen that of the fourteen genuine modes only five are infra-red active, namely those of species B_{1u}, B_{2u} and B_{3u}, because only these species include one of the translations T_x, T_y or T_z. There are two modes with transition moments along *x*, two along *y* and one along *z*. The vibrational modes are shown in Fig. 3.2, and the five infra-red-active modes, which are of special interest, are indicated by the frequencies of the observed absorption bands and the directions of the transition moments corresponding to T_x, T_y and T_z. Obviously in this case the direction of the transition moment could be inferred by inspection.

The spectrum of the oriented polymer is shown in Fig. 3.4 with traces taken with the **E** vector of the radiation parallel and perpendicular to the fibre axis of the polymer, along which the polymer chains tend to lie. In a fibre there is orientation of the polymer chains with respect to this one axis only, and so bands with transition moments along the *x* or the *y* axes of the polymer chains will appear in the spectrum when the **E** vector is perpendicular to the fibre axis. The four fundamental bands of this type are easily identified in Fig. 3.4. They include the strong CH stretching modes at 2851 and 2919 cm^{-1}, a band at 1467 cm^{-1} in which the angle between the CH bonds of each CH_2 group is deformed during the vibration, and the so-called CH_2 rocking mode at *c.* 720 cm^{-1} to which reference has been made earlier in this chapter. This band appears as a rather close doublet (components 720 and 731 cm^{-1}) in crystalline polyethylene samples, but is single when the polymer is amorphous. The origin of the splitting in the crystalline form is considered later. The remaining fundamental mode is one of species B_{1u} with transition moment along the chain axis (wagging mode), a weak band whose identification has been the subject of some disagreement (see Krimm, 1960, for a discussion of this and related matters). Because of the presence of combination bands and also of bands which are associated with the amorphous phase of polyethylene, the assignment of this B_{1u} mode (which gives only weak absorption) has become reasonably certain only by considering the infra-red and Raman bands of a long series of normal paraffins as well as of polyethylene. It appears to have a frequency *c.* 1176 cm^{-1} in the crystalline polymer (Zbinden, 1964; Schachtschneider and Snyder, 1964). The spectrum of thick specimens of polyethylene has a number of

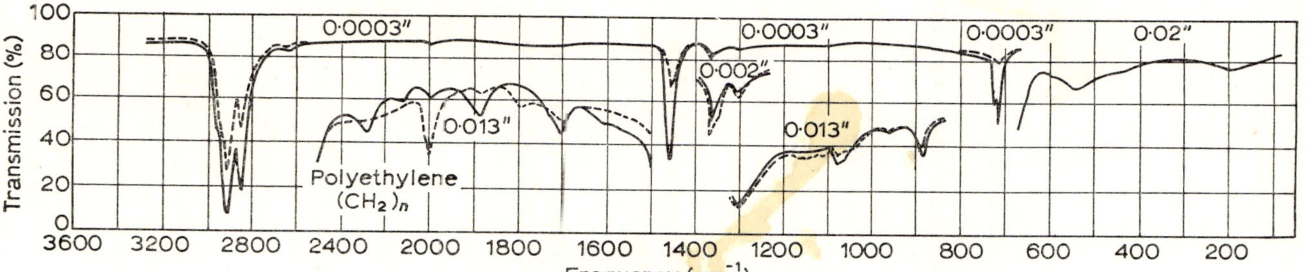

Fig. 3.4 Infra-red spectrum of oriented polyethylene. ——, radiation with electric vector perpendicular to the stretching direction; – – –, radiation with electric vector parallel to the stretching direction. (Krimm, Liang and Sutherland, 1956)

weak bands. Their interpretation was made difficult because the earlier observations were made with polymers produced by the high-pressure process, which are branched, and which have in consequence a large number of terminal CH_3 groups. The linear polyethylenes now available have removed this source of confusion.

7 Infra-red activity of overtone and combination bands

For molecules in which the restoring force is not directly proportional to the displacement, that is to say for anharmonic vibrations, other transitions are possible than those in which the vibrational quantum number for one mode (say the *i*th) changes by ± 1. If the transition is from the ground state to $v_i = 2, 3, 4$, etc., the bands are known as the first, second, third, etc., overtones of ν_i, the changes taking place within one vibrational mode. If the vibrational quantum numbers of two (or more) modes change simultaneously, the transition gives rise to a combination band. Overtone and combination modes are usually weak, and for observation in polymers may require a specimen thickness of some tenths of a millimetre. It is often difficult to analyse such a spectrum in terms of molecular vibrations, partly because inactive fundamental vibrations can contribute to active overtone and combination bands. The frequencies of infra-red inactive modes in polymers are in most cases not known, the Raman effect, from which such information may often be derived, being difficult to observe. The infra-red overtone and combination bands of a bundle of fibres (immersed in a suitable liquid to reduce reflection losses) may be easily observed.

The infra-red activity of these bands is found in essentially the same way as that of fundamentals, by ascertaining whether a component of the dipole moment **M** belongs to the same species as the product $\psi_v'\psi_v''$. The lower state for overtone bands is the ground state ($v'' = 0$) and ψ_0'' is totally symmetric, hence infra-red activity requires one component of **M** to have the same species as ψ_v'. For non-degenerate vibrations, the eigenfunction for any particular normal mode, say ν_i, is symmetric with respect to a given symmetry operation if the vibration ν_i is symmetric for that operation. If ν_i is antisymmetric, then ψ_i is symmetric for even and anti-symmetric for odd values of v.

From this it follows that for overtone bands, if v' is even, the product $\psi_v'\psi_v''$ is totally symmetric, whereas if v' is odd, the product has the same symmetry properties as in the fundamental $0 \rightleftarrows 1$ transition. For non-degenerate bands the activity is found from tables of symmetry species such as Table 3.1. It is evident from this table that polyethylene has no infra-red active first overtones, since no component of **M** belongs to the totally symmetric species A_g.

The only combination bands considered here are those for which the lower state is the ground state with v'' equal to zero. These are sometimes called summation bands, because the frequency is approximately the sum of the frequencies of two or more fundamentals or overtones

$$\nu = v_i\nu_i + v_j\nu_j + v_k\nu_k + \ldots$$

A summation band (for non-degenerate vibrations) will have infra-red activity if one of the components of the dipole moment belongs to the same species as the eigenfunction for the upper state. This will be symmetric or antisymmetric with respect to a certain symmetry element, depending on whether the sum of the vibrational quantum numbers of all vibrations which are antisymmetric for that element is even or odd.

The fact that inactive fundamentals can contribute to overtone and combination bands makes the spectrum complicated. In polyethylene, they have to a great extent been classified, but for this polymer Raman data were obtained (Nielsen and Woollett, 1957).

B CRYSTALLINE ARRANGEMENTS OF POLYMER CHAINS

Many polymers under suitable conditions form ordered regions of limited extent, capable of giving fairly sharp x-ray reflections and deserving the name crystallites. (The degree of order may sometimes be less than would at first sight appear from the diffraction pattern and in some cases only a statistical arrangement, akin to a mixed crystal, exists. Such statistical arrangements are not considered here.) If the molecular attractions between polymer chains in a crystallite are small enough, then the spectrum is not appreciably different from that of isolated polymer chains. If there is only one chain in the unit cell of the crystallite, the infra-red and Raman spectra are almost identical with those of the isolated chain, at least to the extent that the crystalline regions may be considered infinite. The normal modes of vibration of an infinite crystal which are potentially infra-red or Raman active are those in which identical motions occur simultaneously in all unit cells of the lattice (a similar rule has already been described for the one-dimensional lattice of an extended polymer chain). Hence the infra-red or Raman spectrum of an infinite crystal can be predicted by considering the motions of the atoms within the unit cell. The number of potentially active normal modes will be equal to the number for the isolated chain multiplied by the number of chains passing through the unit cell. Only the three translational modes will be of zero frequency. Rotation around the chain axis, which for the isolated chain is of zero frequency, corresponds in the crystal to a rotational lattice mode. (In a lattice mode the internal forces of a molecule are not operative; the restoring forces during the vibration are those imposed by the surrounding molecules in the lattice. Lattice modes can be of translational or of rotational type.) When there are two chains in the unit cell, the vibrations in one chain may be either in phase or out of phase with those in the other chain, and the number of normal modes is twice that of the isolated chain. If the interaction between the chains is from the weak van der Waals forces, the frequency difference between in-phase and out-of-phase modes will be small, and is in fact in many cases unobserved. It is among bands of low frequency (where the force constants which determine the frequency are small) that the splitting of bands from these interactions is most likely to be observed. However, not all the modes which arise in this way will be active. It may be possible to conclude from inspection that certain modes will be inactive, but it must be remembered that in general the symmetry of a chain is lowered when it is put into a three-dimensional lattice, since the force field from neighbouring chains will usually be of different symmetry from that of an individual chain, and this may result in an inactive band becoming active in a crystal. The number of vibrational species and their infra-red activity can be examined by methods analogous to those described above for individual chains. The infinite lattice has symmetry elements which constitute a space group, of which the symmetry elements of the unit cell together with the unit translation (considered as an 'identity' operation) form a sub-group—the unit cell group. From the multiplication table for this group the isomorphous point group is identified and so the symmetry of the various species is found.

The relation between the species of the line and space groups may be found from the following considerations.

Some of the symmetry operations of the line group are found in the space group also. Since the vibrations of the individual chains are essentially unaltered in the crystal (but have specific phase relations with those of other chains), it follows that a vibration of the line group must behave in the same way with respect to these common symmetry operations as do those of the space group. Hence the characters (symmetry) for these operations are looked up in tables of the symmetry species for the line and for the space group, and each line group species is correlated with the space group species having the same characters for the common symmetry elements.

As an example, the relation between the spectrum of crystalline polyethylene and the isolated extended chain will be considered. The spectrum of crystalline polyethylene has been examined in detail by Nielsen and Woollett (1957), Krimm (1960), and Nielsen and Holland (1961). Polyethylene crystals are orthorhombic and have two chains passing through the unit cell (Bunn, 1939). The symmetry elements of the unit cell are shown in Fig. 3.5. It happens by chance that the space group of the unit cell is isomorphous with the same point group D_{2h} which is isomorphous with the line group of polyethylene. The elements common to line and space groups in this case are E, $\sigma(xy)$, $C_2^s(z)$ and i. From Table 3.1 it is seen that for the line group mode the characters for these operations are $+1$, -1, -1, $+1$ for the symmetry types B_{3g} and B_{2g}. Hence in the crystal, the single modes of the B_{3g} and B_{2g} species of the line group each splits into two modes of species B_{3g} and B_{2g}. The splitting of the

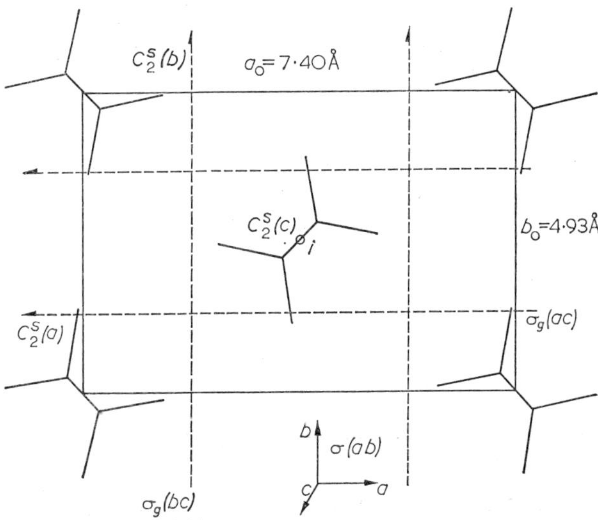

Fig. 3.5 Unit cell and symmetry elements in polyethylene. (After Bunn, 1939)

modes of other symmetry species may be found in the same way; the correlation (for polyethylene) is shown in Table 3.5.

Two consequences of the doubling of the number of normal modes (for two chains in the unit cell) may be mentioned at this stage. First, the three zero-frequency translational modes of the isolated polymer chain each split into one zero-frequency mode and one translational lattice mode of low frequency, whereas the one zero-frequency rotational mode produces two rotational lattice modes. Second, since each mode of the line group splits into two modes which belong to different symmetry species, it is possible for a mode which is inactive in the isolated chain to become active in the crystal (we have seen earlier the physical reason for this). This happens in polyethylene, when the single inactive vibration of species

Table 3.4 Line group and space group fundamental vibrations of polyethylene (from Zbinden, 1964)

Line group				Space group					
		Activity*			Activity		Observed spectra, frequency (cm^{-1})		
Species	Characteristic vibration	IR	Raman	Species	IR	Raman	IR	Raman	Remarks
A_g	C–C stretching (C_3 bending) perpendicular to chain	i	p	A_g	i	p		1061	~1070 computed value
				$B^1{}_g$	i	d			
	CH_2 symmetric stretching	i	p	A_g	i	p		2848	
				B_{1g}	i	d			
	CH_2 bending	i	p	A_g	i	p			Raman line expected around 1640 cm^{-1}
				B_{1g}	i	d			
A_u	CH_2 twisting	i	i	A_u	i	i			For finite chain with polar end group progression band observed 1180–1322
				B_{1u}	M_c	i	1306†		
B_{1g}	Rotation around chain axis			A_g	i	d			Rotational lattice vibration, very low frequency expected
				B_{1g}	i	d			
	CH_2 rocking	i	d	A_g	i	p		1168	
				B_{1g}	i	d			
	CH_2 asymmetric stretching	i	d	A_g	i	p		2883	
				B_{1g}	i	d			
B_{1u}	Translation T_z			A_u		i			Very low frequency expected for translational lattice vibration
				B_{1u}	T_c				Zero frequency for pure translation
	CH_2 wagging	M_z		A_u	i				
				B_{1u}	M_c	i	1176 (1367†, 1352†)		

B_{2g}	C–C stretching	i	d	B_{2g}	i	d		1131	~1150 computed value
				B_{3g}	i	d			
	CH_2 wagging	i	d	B_{2g}	i	d		1415	Doubtful (Snyder 1967a)
				B_{3g}	i	d			
B_{2u}	Translation T_y			B_{2u}	T_b				Zero frequency for pure translation
				B_{3u}	M_a	i			Very low frequency for translational lattice vibration
	CH_2 rocking	M_y	i	B_{2u}	M_b	i	720		
				B_{3u}	M_a	i	731		
	CH_2 asymmetric stretching	M_y	i	B_{2u}	M_b	i	2919		Split in *n*-paraffin with components at 2824 and 2899(?)
				B_{3u}	M_a	i			
B_{3g}	CH_2 twisting	i	d	B_{2g}	i	d		1295	See remark for A_u twisting vibration
				B_{3g}	i	d			
B_{3u}	Translation T_x			B_{2u}	M_b	i			Very low frequency for translational lattice vibration
				B_{3u}	T_a				Zero frequency for pure translation
	CH_2 symmetric stretching	M_x	i	B_{2u}	M_b	i	2851		Split in *n*-paraffin with components at 2850 and 2857
				B_{3u}	M_a	i			
	CH_2 bending	M_x	i	B_{2u}	M_b	i	1463		
				B_{3u}	M_a	i	1473		

* Instead of listing the components of the polarizability tensor, it is shown whether the light in the Raman line is polarized (p) or depolarized (d).

† Observed for amorphous polymer.

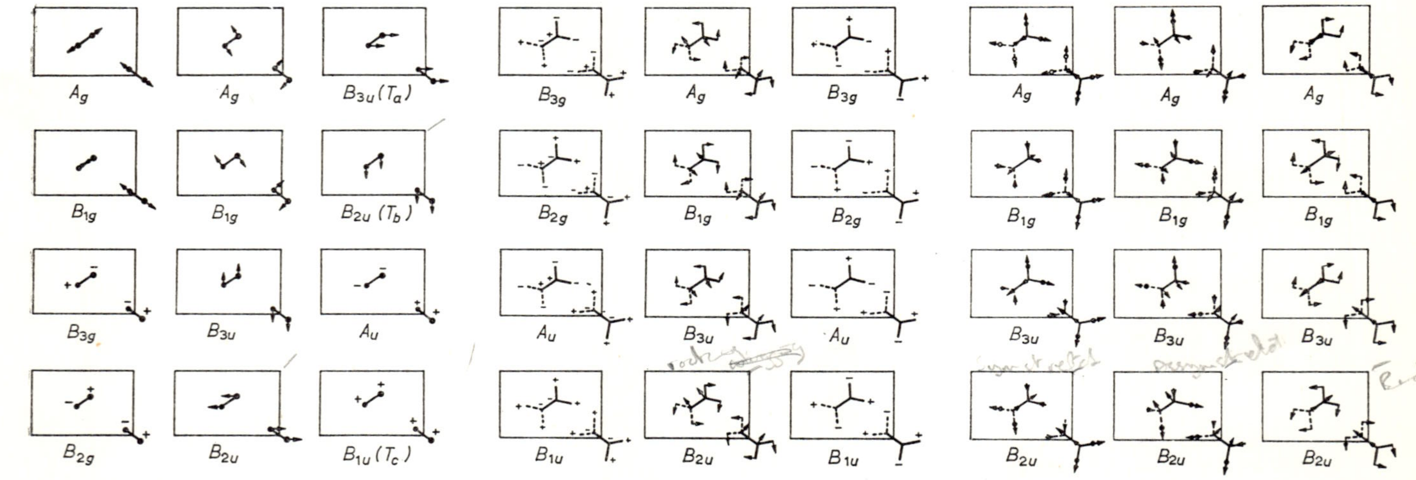

Fig. 3.6 Symmetry modes of polyethylene unit cell. (Krimm, Liang and Sutherland, 1956; the designation has been altered to conform to Table 3.4.)

A_u of the line group (a CH_2 twisting mode) gives rise to an A_u and a B_{1u} mode in the crystal. The former is infra-red inactive, the latter active with transition moment along the *c* axis of the crystal (the polymer-chain axis). The active band is weak; it appears to be the 1050 cm^{-1} observed in the crystalline polymer (Krimm, 1960).

The vibrations in crystalline polyethylene are shown in Fig. 3.6 (Krimm, Liang and Sutherland, 1956), and the line and space group fundamentals are given in Table 3.4. The most striking example of the splitting of the fundamental vibrations when crystallization of the polymer occurs is shown by the behaviour of the CH_2 rocking mode at *c.* 720 cm^{-1}. This band is known to be single in molten polyethylene and double in the crystalline form (Thompson and Torkington, 1945). The splitting into two components (both of which have the **E** vector perpendicular to the chain axis in crystalline, oriented polyethylene (Elliott, Ambrose and Temple 1948b))

Table 3.5 Corresponding species of the line group and space group of polyethylene

Line group		Space group	Line group		Space group
A_g	$\longrightarrow$	A_g, B_{1g}	B_{2g}	$\longrightarrow$	B_{2g}, B_{3g}
A_u		A_u, B_{1u}	B_{2u}		B_{2u}, B_{3u}
B_{1g}		B_{1g}, A_g	B_{3g}		B_{3g}, B_{2g}
B_{1u}		B_{1u}, A_u	B_{3u}		B_{3u}, B_{2u}

was also observed with single crystals of a long-chain hydrocarbon (Krimm, 1954). In this case it was possible to show that the transition moments of the B_{3u} (731 cm^{-1}) and B_{2u} (720 cm^{-1}) modes were along the *a* and *b* axes of the crystal, giving striking confirmation of the theoretical prediction.

The CH_2 deformation mode is also found to be double in crystalline polyethylene. The components are at 1463 and 1473 cm^{-1}; they are not resolved in Fig. 3.5. It is uncertain whether the CH_2 stretching bands (symmetric stretching 2851; asymmetric stretching 2919 cm^{-1}) have been resolved into doublets.

The CH_2 wagging mode has already been mentioned. This very weak band (1176 cm^{-1}) is the only fundamental mode with transition moment along the chain axis. Reference to Table 3.5 shows that this mode (of species B_{1u}) is expected to give B_{1u} and A_u modes in the crystal. The latter is inactive (Table 3.1) and so the wagging mode remains single in the crystalline form (though a change of frequency as compared with the isolated molecule would be expected). It is obvious that bands with a transition moment along the chain axis will become inactive when the two chains in the unit cell are vibrating out of phase.

More detailed consideration of the effect of molecular interaction on vibrations in polyethylene, which is beyond the scope of this book, may be found in papers by Tasumi and Shimanouchi (1965) and Tasumi and Krimm (1967, 1968). The difficult subject of the weak bands in solid polyethylene ascribed to a small proportion of non-planar hydrocarbon chains has been considered by Snyder (1967b).

C FREQUENCIES OF BANDS IN POLYMER SPECTRA

As in simple molecules, the vibrations of polymeric chains may often be identified with those of small chemical groups, and correlations such as those given by Colthup (1950) or in more detail by Bellamy (1958) may be made. The vibrations of pendant groups such as NH, OH, C=O, CH_2, CH_3, etc., may often be identified. However, such vibrations are by no means always isolated in this way, as has sometimes been revealed by observations on polymers in which some of the hydrogen atoms have been replaced by deuterium. In some cases it is found that this affects much more of the spectrum than those bands which might have been attributed to vibrations within the groups where replacement of hydrogen by deuterium has occurred.

In recent years methods for calculating the frequencies of normal modes in chain molecules (in general, helical molecules) have been developed and applied to some of the simpler polymers. For these calculations, force constants derived from analysis of the spectra of simpler molecules are used, preferably from a series of such molecules closely related to the polymer (Schachtschneider and Snyder, 1964). The methods of calculation are described, for example, by Higgs (1953), Tadokoro (1960) and Miyazawa (1964).

CHAPTER 4

Uses of Polarized Radiation

A INTRODUCTION

The early literature on infra-red spectroscopy contains many examples of dichroic spectra, that is absorption spectra which change when the direction of the electric vector of the incident radiation is altered (Schaeffer and Matossi, 1928). These are almost all spectra of inorganic crystals, and were observed with the polarized beam reflected from a mirror of amorphous selenium (Pfund, 1906). Shortly before the Second World War, Ellis and co-workers observed dichroic effects in oriented polymers such as protein fibres. These observations were made with a calcite prism and were restricted to the region near 2 μ, where overtone and combination bands occur. As shown in Chapter 3, the interpretation of these bands can be difficult. The first significant observations of fundamental absorption bands in oriented polymers with polarized radiation were made by Elliott and Ambrose (1947) and Elliott, Ambrose and Temple (1948b) using the newly developed transmission polarizer with selenium films. Subsequently, many applications of this and of the similar silver chloride polarizer (Newman and Halford, 1948; Wright, 1948) to problems of polymer structure have been made.

In single crystals there is a precise relation between the several molecules in one unit cell, and the unit cells are all related to each other by the lattice translations. In so-called crystalline polymers, the repeating units in the polymer chains are related to each other by the unit translations of a three-dimensional lattice, but the dimensions of the polymer within which these relations hold are small. This distance along the chain axis may sometimes be hundreds of Ångstrom units, but regularity in directions perpendicular to this axis is usually restricted to much smaller distances. Even within the regularly arranged region (the 'crystallite') the degree of order may be much less than in single crystals. Outside these regular or fairly regular regions the chains are supposed to be more or less disordered, and in some polymers there may be regions in which no order can be detected, and so they are described as 'amorphous'. Since the length of polymer chains may be much greater than the *c*-axis dimension of a crystallite, it is supposed that a single chain may run through several crystallites, separated by amorphous regions. Where the crystalline portions or micelles merge into the amorphous regions, the molecular chains form fringes at the boundaries, and so this description of the structure of a polymer is known as the 'fringed micelle theory'. The history of the development of this and of other theories has been reviewed by Hearle (1963). The theory gives a reasonable explanation of the properties of cellulose and of many synthetic fibres, including the features of the x-ray diffraction pattern (finite width of the reflections and presence of 'amorphous' rings), the effect of moisture, uptake of dyes, differences in density of the material in bulk and as determined from the size of the unit cell, etc. Evidence from infra-red spectroscopy will be given in this and in the next chapter.

For interpreting the infra-red dichroic effects and for relating them to polymer structure, use will be made of a simplified form of the fringed micelle model. However, it seems unlikely that this corresponds to the real structure in all cases. For example, some fibrous polymers in living organisms appear to be built of a large number of identical units, and so have a quite different structure from that of a polymer in which there is a range of molecular weights. Again, Hosemann (1962) particularly has developed the idea of 'paracrystals' in polymer, and has shown that some of the features in x-ray diffraction patterns attributed to amorphous regions can be accounted for by distortions of the lattice, without postulating the existence of amorphous polymer. This, of course, does not disprove the existence of such regions.

B ORIENTATION IN POLYMERS

The unit cell of the polymer in the ordered parts can belong to any of the crystal forms, and hexagonal, tetragonal, orthorhombic, monoclinic and triclinic forms occur. The dimensions of these ordered regions are much smaller than the wavelength of infra-red radiation, in consequence of which the polymer can behave as if it were optically homogeneous even though it contains regions of very differing order. Polymers in which there is some preferred direction for the chains are birefringent, but the kind of birefringence (uniaxial or biaxial) does not depend on the symmetry of the crystallites, but on their arrangement.

When a polymer is stretched or rolled there is a tendency for the molecules to align along the direction of stretching. Two main types of order may be distinguished.

(*a*) In 'fibre' or uniaxial orientation only *one* axis in the molecules (usually the chain axis, but at least one exception is known, see below) is aligned or partially aligned, with random orientation of the other two axes round the fibre axis. If there are crystallites, then there is orientation of only one crystal axis, usually the one which coincides with the chain axis. The preferred direction within the polymer is the *fibre axis*.

(*b*) If 'double orientation' is present, then *two* axes of the chains and of the crystallites have a preferred direction. This is often the case in polymers whose molecules form layers and which have been pressed or rolled into sheets. Only if the crystallites are orthorhombic will this imply that the third crystal axis has a single preferred direction (in general, there will be two preferred directions for the third crystal axis).

Assuming that appropriate mechanical treatment has been given to induce orientation, the type of orientation will depend on the symmetry of the polymer chains and on the forces which hold the chains together. Molecules with cylindrical or near-cylindrical symmetry such as helices with more than two units in a pitch length will generally have fibre-type orientation. Even planar molecules may orient in this way unless the binding forces tend to produce molecular sheets. When there are strong forces such as hydrogen bonds forming planar molecules into sheets, double orientation is to be expected. This happens in polyamides, for example, when an unoriented fibre is rolled, or when an oriented fibre is pressed.

In (*a*) and (*b*) above, the chain axes will not all lie along the fibre axis, but will be distributed about it. If a direction in space is defined with respect to the fibre axis (z axis) by the angles α and ϕ (Fig. 4.1),

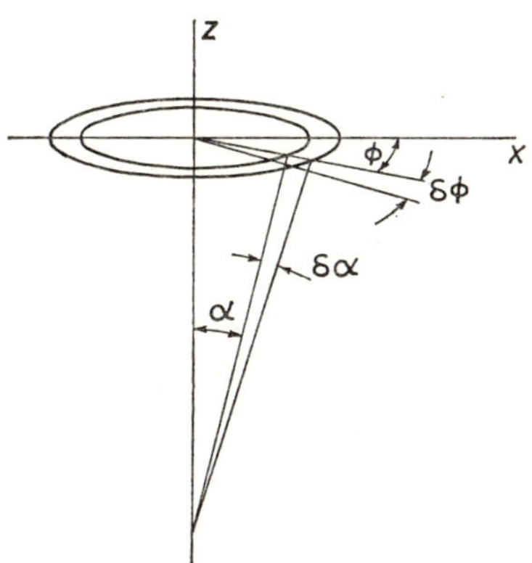

Fig. 4.1 Illustrating the distribution function $a(\alpha, \phi)$ (see text).

the number of molecules with chain axes which lie between $\phi + \delta\phi$, $\alpha + \delta\alpha$ is given by a distribution function $a(\alpha, \phi)$ and is equal to $a(\alpha, \phi)\delta\alpha\delta\phi$. The distribution may be shown by a polar diagram, with a vector whose length in the direction α, ϕ is proportional to $a(\alpha, \phi)$. Figure 4.2 shows several polar diagrams of this kind, corresponding to conditions which occur most frequently in polymers. It is to be noted that this figure refers to the orientation of the chain axis only. A more complete discussion of the types of orientation in polymers is given by Zbinden (1964).

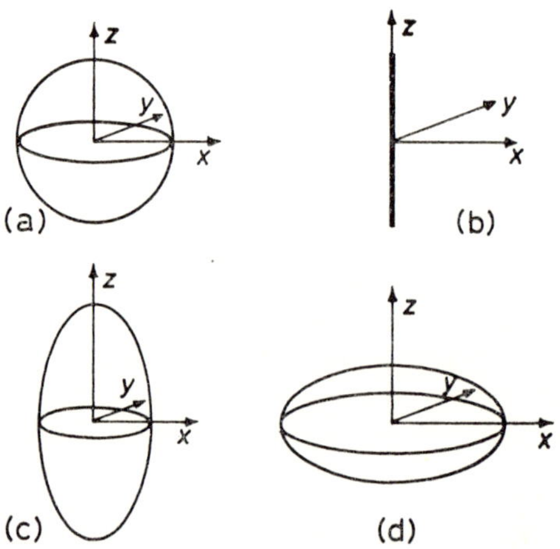

Fig. 4.2 Polar distribution diagrams. The radius vector has a length proportional to the distribution function $a(\alpha, \phi)$. (**a**) Random orientation; (**b**) perfect alignment along one axis; (**c**) partial (fibre-type) orientation; (**d**) partial planar orientation.

The partial uniplanar orientation shown in Fig. 4.2d is very common in unstretched polymer films which have been cast from solution. Such films do not show birefringence if laid flat on the stage of a polarizing microscope, but if tilted they will appear birefringent. When there is fibre-type orientation and the distribution function has cylindrical symmetry, the polymer is optically uniaxial. Double orientation produces a polymer which is optically biaxial (having three different principal refractive indices).

If we take the fringed micelle model for polymers in general, two distribution functions will be needed, one for crystallites and the other for the amorphous regions (assuming that the regions of intermediate order can be left out of account). For many purposes, semiquantitative results are still of considerable value, and much may be done with a simple model in which the amorphous regions are taken as completely random, and crystallites as perfectly oriented (Fraser, 1953b). The dichroic effects corresponding to a variety of distribution functions have been examined (Brown and Trotter in Bamford, Elliott and Hanby, 1956; Beer, 1956; Zbinden, 1964; Fraser, 1953b, 1956, 1958a, 1958b).

C DICHROISM IN POLYMERS

As explained in Chapter 3, the intensity of an absorption band depends on the transition moment for the corresponding vibrational transition, which is a vector. If all directions of the electric vector of the radiation are equally probable, the integrated absorption coefficient is

$$\int_{-\alpha}^{\alpha} k(\nu)d\nu = \frac{8\pi^2\nu_i N}{3ch} M^2 \tag{4.1}$$

Here $k(\nu)$ is the absorption coefficient for a frequency ν, N is the number of absorbing centres per cm^3, ν_i is the frequency of a normal mode which for fundamental transitions is the frequency of the radiation absorbed, and M is the magnitude of the corresponding transition moment (see p. 46). When polarized radiation is employed, the changing electric vector of the radiation **E** will be inclined to the transition moment vector **M** at some angle θ, so the magnitude of the component of **M** in the direction of **E** will be $M \cos \theta$. The rate of absorption of energy will now be proportional to $M^2E^2 \cos^2 \theta$, that is to $(\mathbf{P} \cdot \mathbf{E})^2$.

Since we are here concerned only with the way in which the absorption depends on the angle θ, it is convenient to define a quantity P_E such that the rate of absorption per absorber is

$$\delta I = -P_E{}^2 I \tag{4.2}$$

where I is the intensity of the incident radiation of the frequency absorbed. Clearly P_E is proportional to the component of the transition moment in the direction of the **E** vector of the radiation. If the N absorbers are independent, and the radiation falls normally on a thin sheet of thickness δx, the rate of absorption of energy per cm^2 is

$$\delta I = -NP_E{}^2 I \delta x \tag{4.3}$$

whence

$$I = I_0 e^{-NP_E{}^2 x} \tag{4.4}$$

where I_0 and I are the incident and transmitted intensities, and the optical density of a sheet of thickness x is

$$\log_{10} I_0/I = NP_E{}^2\, x \log_{10} e \tag{4.5}$$

From this expression we see that if polarized radiation were incident on a sample in which the transition moments of all absorbers (for a given absorption band) were parallel, the optical density would

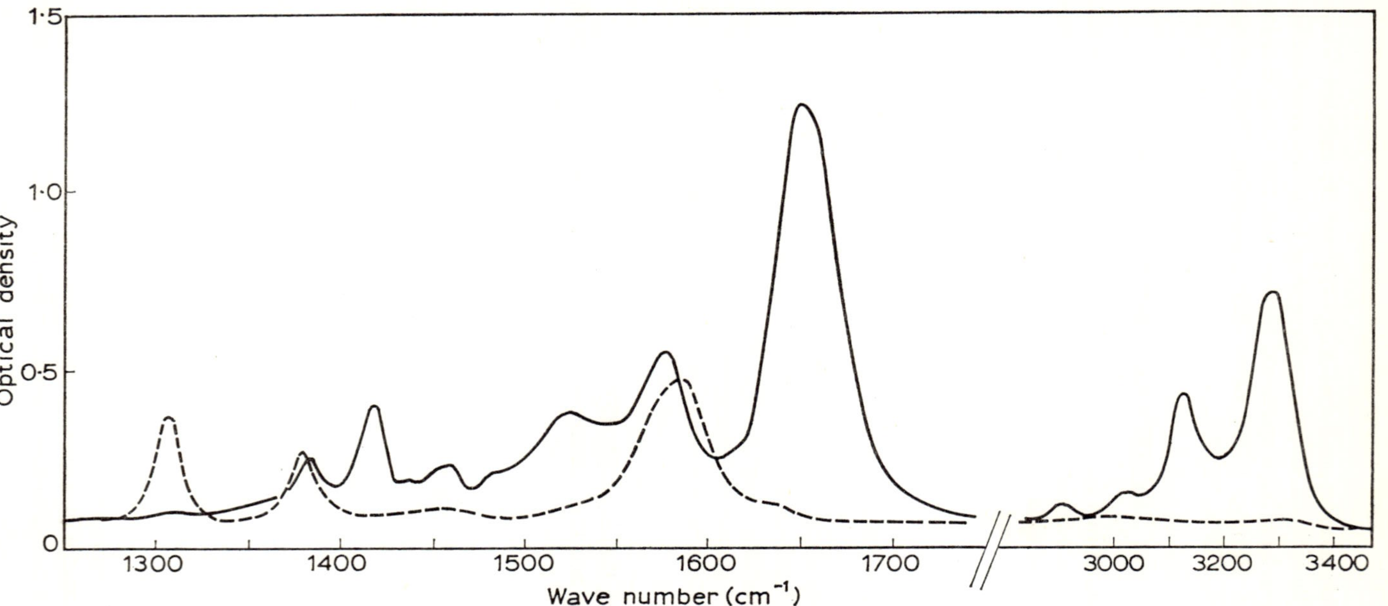

Fig. 4.3 Dichroism in the spectrum of crystalline N-methyl acetamide at −100° C. ——, **E** vector parallel to *a* axis; – – –, **E** vector parallel to *c* axis. (Bradbury and Elliott, 1963b)

vary as $\cos^2 \theta$, being zero when there is no component of the transition moment in the direction of the **E** vector of the radiation. This can happen in crystal specimens of small molecules, as shown in Fig. 4.3. In oriented polymers, the absorbers do not have transition moments all in the same direction. Even when this direction coincides with the chain axis, the distribution of chain directions round the fibre axis reduces the dichroic effect.

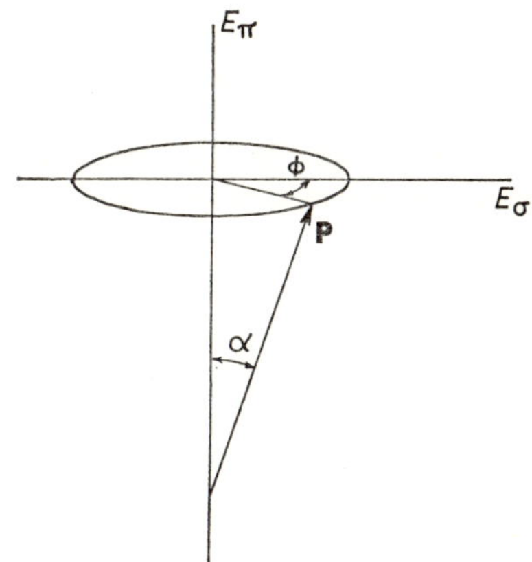

Fig. 4.4 Direction of the transition moment **P** in a fibre.

PERFECTLY ORIENTED CHAINS IN A FIBRE

Figure 4.4 shows the direction of the transition moment **P** for one absorbing group, making an angle α with the fibre axis and in a plane making an angle ϕ with some chosen reference plane (the plane normal to the direction of incidence of the radiation). If the electric vector is set parallel to the fibre axis

$$P_E = P \cos \alpha \tag{4.6}$$

and the corresponding optical density for N absorbers is

$$D_\pi = NP^2x \cos^2 \alpha(\log_{10} e) \tag{4.7}$$

When the polarizer is rotated so that the **E** vector is perpendicular to the fibre axis, the number of absorbers within angles ϕ and $\phi + \delta\phi$ is $Nx \dfrac{\delta\phi}{2\pi}$, and $P_E = P \sin \alpha \cos \phi$. The contribution to the optical density is

$$Nx \frac{\delta\phi}{2\pi} P^2 \sin^2 \alpha \cos^2 \phi(\log_{10} e).$$

Hence

$$\begin{aligned} D_\sigma &= \frac{1}{2\pi} \int_0^{2\pi} NxP^2 \sin^2 \alpha \cos^2 \phi(\log_{10} e)d\phi \\ &= \tfrac{1}{2}NP^2x \sin^2 \alpha(\log_{10} e) \end{aligned} \tag{4.8}$$

The dichroic ratio is

$$D_\pi/D_\sigma = 2\cot^2\alpha \tag{4.9}$$

Evidently if α has the value 54° 45′ the dichroic ratio will be 1, that is the absorption will be independent of the direction of the **E** vector of the radiation.

EFFECT OF UNORIENTED POLYMER

If only a fraction f of the polymer is oriented, and the other fraction $1-f$ is completely unoriented, the latter will absorb as if one-third of its groups had their transition moments along each of the rectangular x, y, z axes in Fig. 4.4. Hence we must add the contribution of these groups to that of the oriented part (Fraser, 1953b). We then have

$$D_\pi = NP^2x\log_{10}e(f\cos^2\alpha + \tfrac{1}{3}(1-f))$$
$$D_\sigma = NP^2x\log_{10}e(\tfrac{1}{2}f\sin^2\alpha + \tfrac{1}{3}(1-f))$$

whence
$$D_\pi/D_\sigma = \frac{f\cos^2\alpha + \frac{1}{3}(1-f)}{\frac{1}{2}f\sin^2\alpha + \frac{1}{3}(1-f)} \tag{4.10}$$

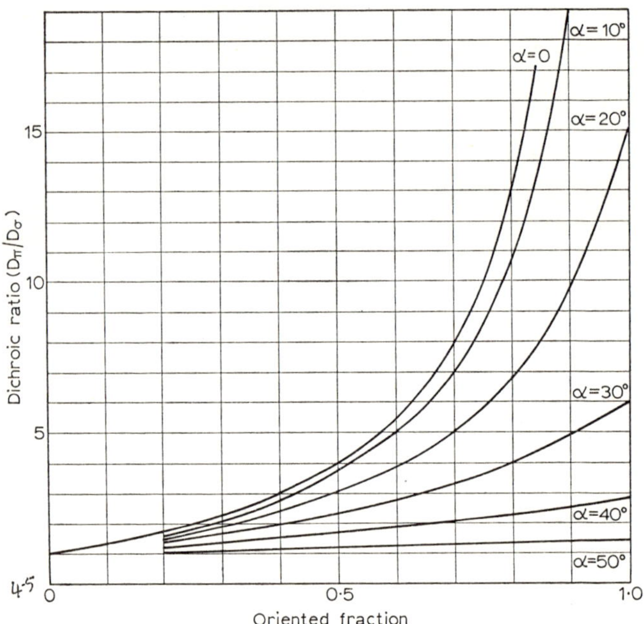

Fig. 4.5 Relation between the dichroic ratio D_π/D_σ and the fraction of perfectly oriented chains in a polymer for various values of α; α is the angle which the transition moment makes with the chain axis. The remainder of the polymer is supposed to be randomly oriented. (Bamford, Elliott and Hanby, 1956)

The model used to derive this expression does not correspond to a real polymer, though for some it may not be very far from reality. Beer (1956) has shown that for fibre-type orientation it is always possible to represent a polymer in this way, so that if a value of f is chosen which predicts the correct dichroic ratio for one value of α, then the dichroism will be correctly predicted for all other values of α. The assumption underlying the above formula is that the magnitude of the transition moment is not dependent on the conformation of the polymer chain. It cannot be true, obviously, for bands which appear only in the crystalline or only in the amorphous parts. It is best approximated by those bands whose frequencies are characteristic of small chemical groups in the molecule, which appear at the same frequency in materials of widely differing composition.

It is convenient to have graphs giving the value of the dichroic ratio as function of the oriented fraction f for a series of values of α, the angle which the transition moment makes with the fibre axis. Such graphs have been constructed by Bamford, Elliott and Hanby (1956) and are reproduced in Figs. 4.5 and 4.6. Beer (1956) has given the same information with f and α as variables, and has discussed how such graphs may be used.

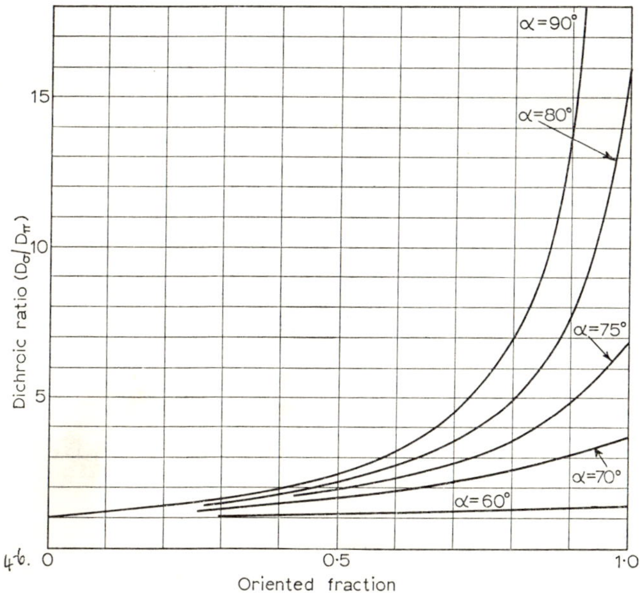

Fig. 4.6 As in Fig. 4.5 except that the more convenient function D_σ/D_π is plotted for values of α between 60 and 90°.

Fraser's equation may be put into a different form

$$f = \frac{(R - 1)(R_0 + 2)}{(R_0 - 1)(R + 2)} \qquad (4.11)$$

in which R is the dichroic ratio for the whole polymer and R_0 the dichroic ratio for the same band in the perfectly oriented part.

AVERAGE ANGLE OF ORIENTATION

A different model for a polymer with fibre-type orientation is got by supposing that all chains make the same angle θ with the fibre axis. If, as before, α is the inclination of the transition moment to the molecular axis, the dichroism is now given by the expression (Fraser, 1953b)

$$D_\pi/D_\sigma = \frac{2 \cot^2 \alpha \cos^2 \theta + \sin^2 \theta}{\cot^2 \alpha \sin^2 \theta + (1 + \cos^2 \theta)/2} \qquad (4.12)$$

or the equivalent expression (Beer, 1956)

$$D_\pi/D_\sigma = \frac{2 \cos^2\alpha + \dfrac{2 \sin^2\theta}{2 - 3 \sin^2\theta}}{\sin^2\alpha + \dfrac{2 \sin^2\theta}{2 - 3 \sin^2\theta}} \qquad (4.12a)$$

This model, like the one described in the preceding section, is valid for fibre-type orientation; that is, the dichroic ratio for all values of α can be correctly predicted if θ has been determined from measurement of the dichroic ratio of a band for which the angle α is known.

It is possible to use a model in which there is a fraction f in which all the chains make an angle θ with the fibre axis, and the remaining fraction $1 - f$ is unoriented. Of the models considered, this is probably closest to the state of some polymeric materials. The presence of a highly oriented fraction is shown in many polymers by the short length of the diffraction arcs of the x-ray diffraction pattern, while other properties suggest that a less oriented fraction is also present. For such a polymer the dichroic ratio is given by

$$D_\pi/D_\sigma = \frac{f \cos^2 \alpha + \dfrac{2 \sin^2 \theta}{2 - 3 \sin^2 \theta} + \frac{1}{3}(1 - f)}{\frac{1}{2} f \sin^2 \alpha + \dfrac{2 \sin^2 \theta}{2 - 3 \sin^2 \theta} + \frac{1}{3}(1 - f)} \qquad (4.13)$$

D TRANSITION MOMENT DIRECTIONS

If we leave out of account for the moment the effect of the association of chains in crystallites, the transition moments in regular chains are either along or normal to the chain axis (see Chapter 3). This is a consequence of the splitting of the vibrational modes of the monomeric groups into as many components as there are groups in the repeat length of the chains. In many spectra these components are not resolved. It is generally assumed that the intensities of the parallel and perpendicular bands are proportional to the squares of the corresponding components of the transition moments of the individual monomer units. If this is so, then, although the assumption made in deriving expressions for dichroic ratios of oriented polymers (that the N oscillators are independent) is not true for regular chains, this does not make the expressions inaccurate.

The direction of the transition moment in the monomer units can sometimes be inferred from symmetry. Thus, the symmetric and asymmetric CH stretching modes in extended polyethylene must have transition moments normal to the chain axis. In polyethylene terephthalate, however, it does not follow that these transition moments in the

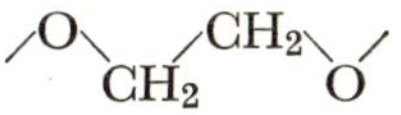

$$\diagup O \diagdown \underset{CH_2}{} \diagup CH_2 \diagdown \underset{O}{} \diagup$$

part of the chain (even in the *trans* form) are simply related to the bisector of the CH_2 group and to the CH_2 plane. The presence of the oxygen groups lowers the symmetry of the CH_2 groups and the direction of the transition moments is not strictly predictable. In polytetrafluorethylene (which forms a regular helix) it can be predicted that for the symmetric CF_2 stretching mode the transition moment will be perpendicular to the helix axis, but that for the asymmetric mode is not.

In order to determine the transition moment direction for some

Table 4.1 Directions of transition moment for the amide bands, measured from the C=O bond direction (positive direction shown by arrow) (from Bradbury and Elliott, 1963b)

Compound	NH str.	Amide I	II	III
N-acetylglycine	+19°(±10)	—	—	—
Acetanilide	+11 or −8°	+22°(±2)	+72 or −57°	—
NN′diacetyl hexamethylene diamine	+8°	+17°	+68, +77° (double)	—
N-methyl acetamide	+8 or +28°	between +15 and +25°	+73 or −37°	+96 or −60°

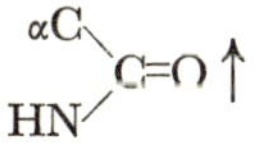

αC\
　C=O ↑
HN/

polymer bands, model compounds of simple substances of known crystal structure have been used. This has been done especially for the vibrations of the mono-substituted amide group because of its importance in proteins, polypeptides and polyamides (see Chapter 5). Table 4.1 gives the directions of the transition moment for the strongest in-plane bands of this group in four crystalline model compounds. In some cases there is an ambiguity, because of the symmetry of the crystal, which is resolved when results for more than one compound are available.

Pendant benzyl groups occur in polystyrene and in the side-chains of some polypeptides. The directions of the transition moments can be inferred from symmetry, but there are many normal modes of similar frequencies, making identification uncertain. Mitsui, Iitaka and Tsuboi (1967) have measured the infra-red spectrum of crystalline dibenzyl in order to identify bands with transition moments along the three principal directions of the benzyl groups. The strong bands at 700 and 755 cm^{-1} have vibrations normal to the plane of the ring. The band at 1490 cm^{-1} has a transition moment along the axis of the phenyl group, and for the band at 1452 cm^{-1} the vibration is in the plane of the ring, perpendicular to the axis.

The direction of the transition moment for the OH stretching vibration has been measured in adipic acid by Ambrose, Elliott and Temple (1951). The crystal structure of this material has been determined using x-ray diffraction by MacGillavry (1941) and by Morrison and Robertson (1949). The position of the hydrogen atoms was not determined by the x-ray investigation, but assuming an angle of 105° between the C=O and OH bonds it appears that the transition moment is along the OH bond.

E EXAMPLES OF THE USE OF POLARIZED RADIATION

The principal applications of polarized radiation in the spectroscopy of oriented polymers are as follows:

(1) to assist in determining chain conformation and structure;
(2) for estimating the degree of orientation;
(3) for the detection of conformation-sensitive bands and of 'crystallinity bands';
(4) as an aid to the assignment of bands to vibrational modes;
(5) to improve the resolution in the spectra of oriented polymers.

1 Chain conformation and structure

The fibrous proteins and the related synthetic polypeptides furnish striking examples of the use of polarized radiation. It was

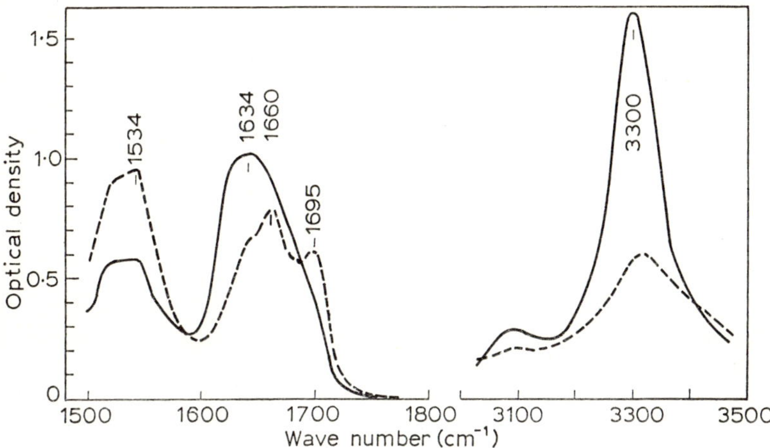

Fig. 4.7 Spectrum of silk gut. ——, **E** vector perpendicular to fibre; – – –, **E** vector parallel to fibre. (Ambrose and Elliott, 1951b)

known (from x-ray work by Astbury *et al.*) that in some fibrous proteins the chain was extended and in others it was folded, but the nature of the fold was not known. The observations of Ambrose, Elliott and Temple (1949b) and Ambrose and Elliott (1951b) showed that, whereas in the extended forms such as silk and feather keratin the NH and C=O stretching vibrations of the polypeptide chains appeared most strongly when the **E** vector of the radiation was perpendicular to the fibre axis, in folded forms such as the α-keratin of hair they were strong in the parallel direction of **E**. It was deduced that in folded forms the NH and C=O groups are more nearly along the chain axis than perpendicular to it. The spectrum of silk is shown in Fig. 4.7, and a more recent spectrum of α-keratin (Bendit, 1966) is shown in Fig. 5.13a. Similar observations by

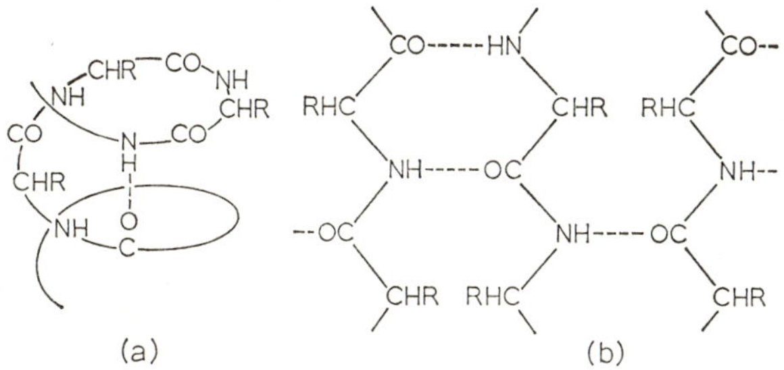

Fig. 4.8 Diagrammatic representation of (**a**) the α-helix, and (**b**) the extended chain or β structure of a polypeptide. (Bamford, Elliott and Hanby, 1956)

Ambrose and Hanby (1949) and Ambrose and Elliott (1951a) showed that folded and extended forms of synthetic polypeptide chains could be detected in the same way. The results of the study of these materials by infra-red spectroscopy and by x-ray diffraction (Bamford, Hanby and Happey, 1951a and b) led Pauling and Corey (1951) to propose that the α-helix fold which they had discovered by building scale models of polypeptide chains was probably the conformation of the chain in these materials. A diagrammatic representation of the extended and α-helix forms of the polypeptide chain is given in Fig. 4.8, which shows how the NH and C=O stretching vibrations will have opposite dichroic character in these two forms. Some infra-red spectra of oriented synthetic polypeptides are shown in Fig. 4.9. These may be compared with the spectrum of a polyamide in Fig. 4.10, when it will be seen that the NH stretching

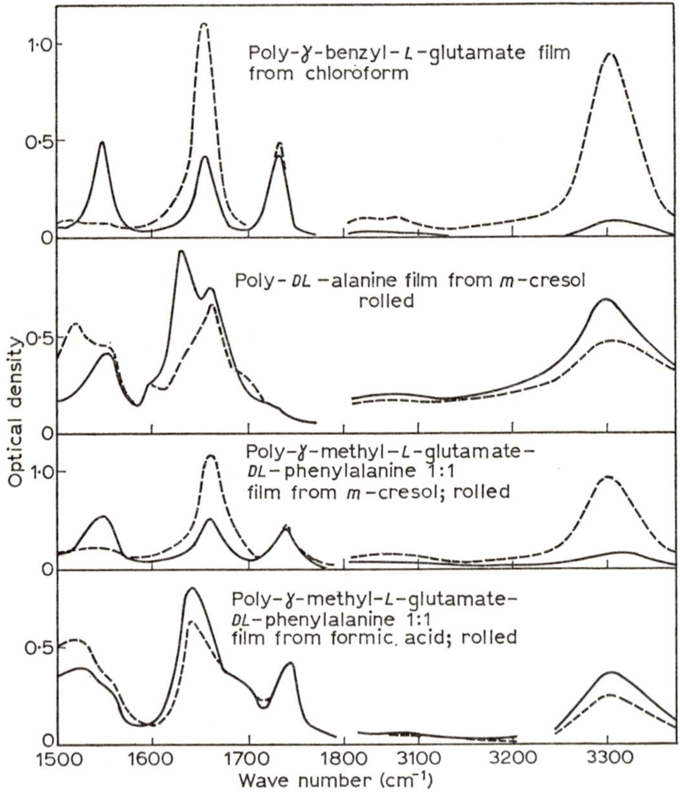

Fig. 4.9 Absorption spectra of oriented films of some synthetic polypeptides. ——, **E** vector perpendicular direction of rolling; – – –, **E** vector parallel to direction of rolling. (Ambrose and Elliott, 1951a, Bamford *et al.*, 1956)

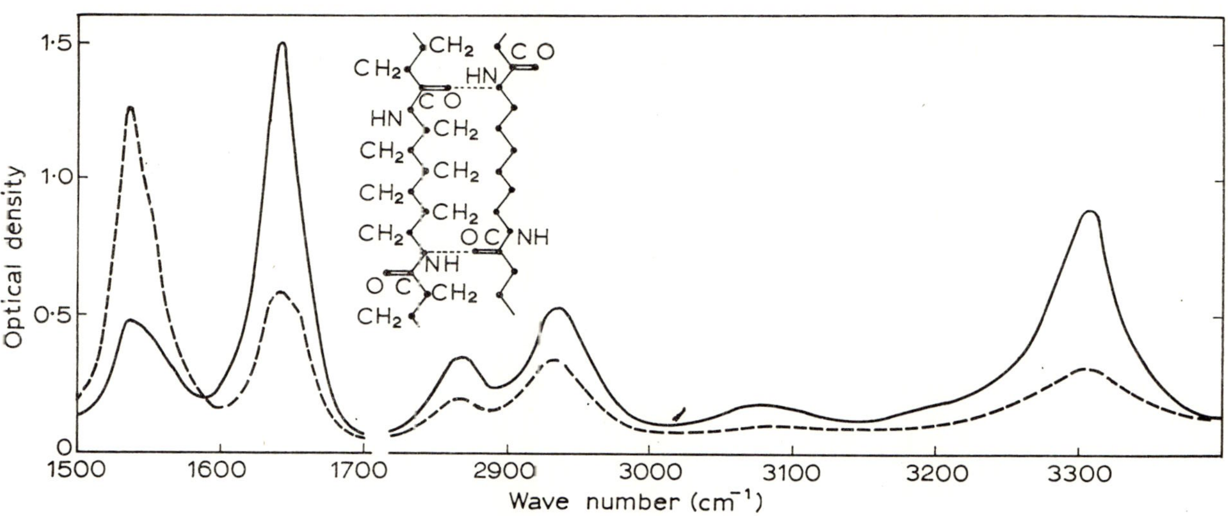

Fig. 4.10 Absorption spectrum of section of drawn nylon 66 monofil. ——, **E** vector perpendicular to fibre axis; – – –, **E** vector parallel to fibre axis. (Bamford, Elliott and Hanby, 1956)

band (Amide A, 3300 cm^{-1}), the C=O stretching band (Amide I, *c.* 1640 cm^{-1}) and the NH deformation band (Amide II, *c.* 1540 cm^{-1}) are similar in the extended or β form of the polypeptide and in the polyamide. It may be noted that the Amide I band has a frequency which is conformation-dependent. In the α-helix form it appears at 1650–1660 cm^{-1}, in the β form at 1630–1640 cm^{-1} (Ambrose and Elliott, 1951a; Elliott, 1953; Bamford, Elliott and Hanby, 1956). The origin of this apparent shift is discussed in Chapter 5. It is really a splitting with a wide separation in the β form, one strong component at 1630–1640 cm^{-1} with perpendicular dichroism, the other much weaker band with parallel dichroism at 1690–1700 cm^{-1}. It may be seen in the spectrum of silk in Fig. 4.7.

A knowledge of the dichroism of collagen in mouse-tail tendon (Fraser, 1950) assisted in rejecting some models of the chain fold in this fibrous protein. As Fig. 4.11 shows, the spectrum is similar in

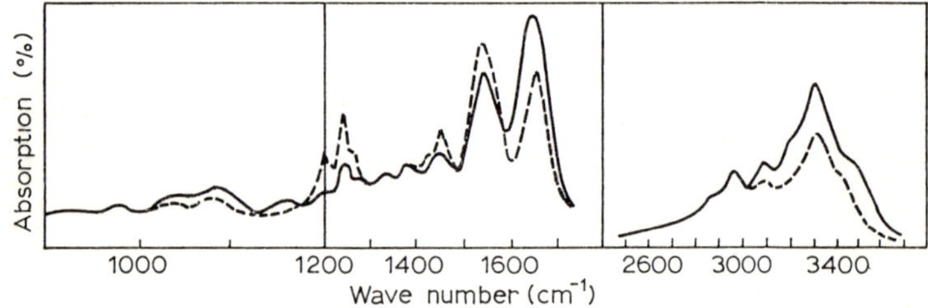

Fig. 4.11 Microspectrometer record of mouse-tail tendon. ——, **E** vector perpendicular to fibre; – – –, **E** vector parallel to fibre. (Fraser, 1950)

character to that of an extended polypeptide. A similar spectrum is obtained from oriented gelatine (Ambrose and Elliott, 1951b). The x-ray diffraction pattern of collagen suggests a peptide translation along the helix axis of 2·86 Å (compared with 3·6 Å in a fully extended chain). The current models for collagen (Ramachandran and Kartha, 1955) have a slightly twisted chain, which retains the dichroic features of the β form of protein. The frequency of the NH stretching mode (3330 cm^{-1}) is that of the *trans* form of the amide group, but the *cis* form (which was at one time proposed in a model for collagen) has a lower frequency.

An interesting variant of the extended (β)-chain form of proteins and polypeptides is known as the 'cross-β' form (Astbury, Dickinson and Bailey, 1935) revealed originally by its unusual x-ray diffraction pattern, in which a reflection corresponding to *c.* 4·7 Å appears on the meridian of the pattern, and so is associated with a corresponding periodicity along the fibre axis. This period is related to the spacing between extended (or near-extended) chains connected by hydrogen

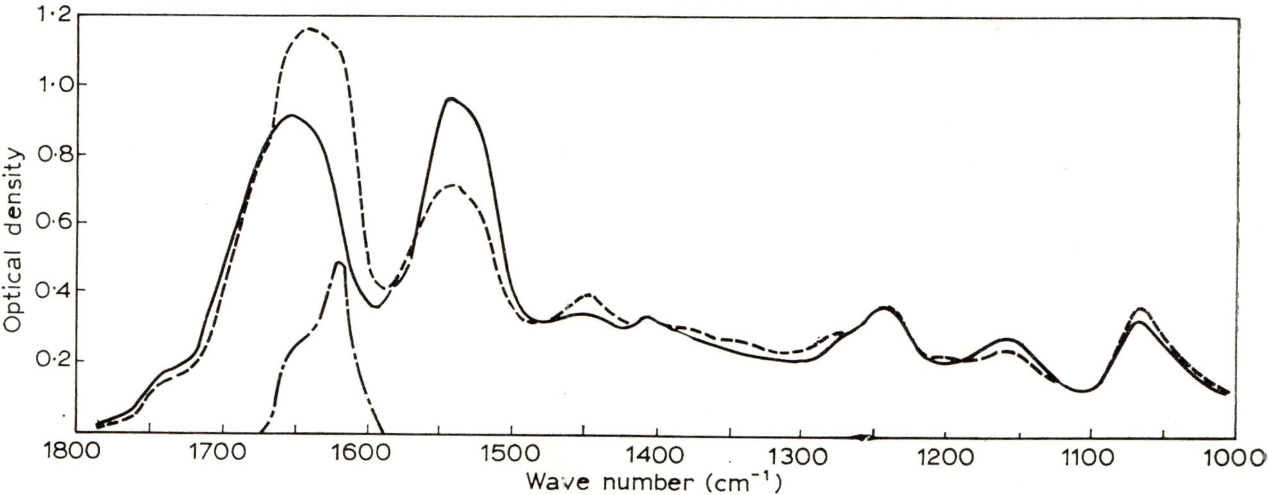

Fig. 4.12 A cross-β spectrum of a section of a fibre bundle in poly-methyl methacrylate. Fibres from the egg-stalk of the green lace-wing fly *Chrysopa flava* (spectrum supplied by Dr K. D. Parker). ——, **E** vector perpendicular to fibre axis; – – –, **E** vector parallel to fibre axis; — · — · — $E_\pi - E_\sigma$, showing the characteristic β-form peak at *c*. 1630 cm^{-1}.

bonds. In the cross-β form, then, the chains are expected to be perpendicular to the fibre axis. This is shown by infra-red spectra with polarized radiation, the NH and C=O stretching vibrations having the strongest absorption when the **E** vector is along the chain axis. At the same time, the frequency of the C=O band (Amide I) is that of the extended form, 1630 cm^{-1}. Many examples of cross-β structures are known; for instance, fibrous insulin (Ambrose and Elliott, 1951c; the silk of the egg-stalk of the green lace-wing fly (Parker and Rudall, 1957), the synthetic polypeptide poly-β-*n*-propyl-L-aspartate (Bradbury *et al.*, 1960), etc. A cross-β spectrum is shown in Fig. 4.12.

Infra-red spectra of doubly oriented polymer films have been used to show how the molecular planes are oriented with respect to the surface of the film. In one method, the film is tilted round an axis in the plane of the film, normal to the fibre axis. If the axis of tilt is arranged parallel to the spectrometer slit, the radiation can make a fairly small angle with the fibre axis (depending on the angle of tilt). It is desirable to place the tilted specimen between prisms with an immersion medium to make optical contact (Ambrose, Elliott and Temple, 1949a) or between hemispheres (Zbinden, 1964). An alternative method is to make a thick, doubly oriented specimen and to cut a thin transverse section with a microtome. This method is more satisfactory than tilting the specimen, since the radiation beam can be made to go along the fibre axis, but it is difficult to cut sufficiently thin transverse sections.

Arimoto (1964) has used the dichroic effects in a doubly oriented (rolled) specimen of nylon 6 to determine the direction of the hydrogen bonds. This polymer has at least two well-differentiated crystal forms, with different chain conformations in each. In the α form (which occurs commonly in melt-spun fibres) the chains form sheets in which there is an alternating sequence of the $(CH_2)_5$ CO·NH units of the chains ('antiparallel sheets'). The chains in this sheet are held together by NH . . . O=C bonds. When nylon 6 in the oriented α form is treated with a solution of iodine in aqueous potassium iodide, the polymer absorbs iodine and swells. On removing the iodine by treatment with aqueous sodium thiosulphate, the polymer is found to be still oriented, but gives an x-ray diffraction pattern different from that of the original α material (Ueda and Kimura, 1958). The x-ray diffraction pattern of the new (γ) form has been studied by several workers (Kinoshita, 1959; Arimoto, 1964; Vogelsong, 1963; Bradbury, Brown, Elliott and Parry, 1965). In Arimoto's work, transverse sections of doubly oriented nylon 6 in the α form were examined with polarized radiation incident normally on the section, the radiation being then parallel to the fibre axis. The dichroism of the NH stretching and C=O stretching

(Amide I) bands showed that these bonds were preferentially oriented in the plane of the rolled sheet. When the specimen was converted to the γ form, the dichroism showed that these bonds (between which the hydrogen bonds are formed) now had a preferred orientation perpendicular to the plane of the rolled sheet. This suggests an explanation of an odd result of the x-ray investigations, namely that in the α form the hydrogen bonds are formed between 'antiparallel' chains (Holmes, Bunn and Smith, 1955), but between 'parallel' chains in the γ form (Kinoshita, 1959). It would seem that the hydrogen bonds are broken by the interposition of iodine atoms (probably with some separation of the chains) and that when the iodine is removed, new hydrogen bonds are formed between chains which in the α form had belonged to two adjacent sheets of antiparallel chains. The new hydrogen bond is formed between chains which are nearest neighbours, and these appear to have their chain sequences in the same sense (hence 'parallel' chains).

Measurements of infra-red dichroism have been used to help in deciding the conformation of the long side-chains in the ω form of poly-β-benzyl-L-aspartate,

$$\left(\begin{matrix}CO\\NH\end{matrix}\!\!>\!CH{\cdot}CH_2COOCH_2C_6H_5\right)_n$$

by Bradbury *et al.* (1962). The ω form is an α-helix in which the structure is distorted into one which packs tetragonally, in which the helices have 4-fold screw symmetry. Interactions between the benzyl groups appear to be the cause of the distortion. The ω form is produced when the α-helix form of the polypeptide is heated to about 160° C and cooled, apparently without loss of orientation. The model in which the fraction f is perfectly oriented, with the fraction $1 - f$ completely disoriented, was used, and it was assumed that the direction of the transition moment of the NH stretching band was known. From the geometry of the α-helix (Pauling and Corey, 1951) and from the direction of the NH transition moment in NN′ diacetyl hexamethylene diamine (Sandeman, 1955) this transition moment is expected to make an angle of about 20° with the helix axis. For a perfectly oriented α-helix the NH band should have a dichroic ratio 15 : 1. A dichroic ratio 14 : 1 has been observed in poly-γ-benzyl-L-glutamate (Ambrose and Elliott, 1951a). Since it is unlikely that this polymer would have been almost perfectly oriented, the angle the transition moment makes with the axis is probably less than 20°. Assuming it is about 18°, the value of f calculated from the observed NH dichroism of a specimen in the α-helix form is 0·85. Assuming further that the oriented fraction in the ω form is the same, the angles α between the transition moment and chain axis can be calculated from the observed dichroic ratios and are given in Table 4.2. These angles do not, of course, define the structure but when used to correct a model are of considerable help. Table 4.2 includes the angles α in a model proposed for the ω structure.

A similar series of measurements has been made on oriented films of poly-γ-benzyl-L-glutamate by Tsuboi (1962).

2 Degree of orientation

If the direction of the transition moment for a certain band is known, it is obvious that the measured dichroic ratio of this band can be used to determine an average angle of orientation of the chain axis, using equation (4.12). This will probably not be physically very meaningful, but it may afford a useful way of classifying

Table 4.2 Frequencies, assignments and dichroic properties of main infra-red absorption bands in poly-β-benzyl-L-asparate films (from Bradbury *et al.*, 1962)

Assignment	Frequencies (cm^{-1})				Dichroic ratio $D\pi/D\sigma$		Angle α between transition moment and chain axis	
	α-helix	α-helix (N-deuterated)	ω form	ω form (N-deuterated)	α-helix	ω form	α-helix	ω form
NH stretch (bonded)	3308		3298		9·0	7·0	18°	22°*
Weak NH stretch	3082		3075					
Aromatic CH stretch	3042	3041	3043	3041	1·0	1·4		
ND stretch		2465 2440		2456				
Ester C=O	1739	1736	1730	1731	1·0	1·5	54	48 (50)
Amide I	1664	1658	1677 1670	1665	2·4	3·0	40	36 (37)
Amide II	1561		1538 1520		0·16	0·20		
Aromatic νC–C	1500	1500	1501	1500	1·0	~0		80 (81½)
Amide II′		1458		1458				
δCH_2 (adjacent to C=O)	1416	1414	1420	1420	7·0	1·3	17½	50½ (35)
Amide III	1308		1299					
Ester C–O	1165	1168	1164	1170	0·38	0·22	69	76 (76½)
Aromatic and CH	752	750	753	753	1·0	0·9		54 (54)
Aromatic ϕC–C	695	697	698	698	1·0	1·1		54 (54)

* The figures in the last column (in parentheses) are from a model of the ω-helix.

oriented material. Alternatively, the model of the fully oriented and fully unoriented fractions may be used, making use of Fig. 4.5 or Fig. 4.6.

Perhaps the best method of characterizing orientation is by combining infra-red methods with x-ray diffraction, if possible on the same specimen. The equatorial reflections are the most suitable, and may be photometered to find at what angle with the equator the peak intensity falls to half-value. This angle may be equated with the average angle of disorientation θ in equation (4.13).

The choice of the band whose dichroism is used to measure the degree of orientation will have an influence on the result. Unless one is intending to estimate the orientation of crystalline and amorphous regions separately, it is obvious that bands which are peculiar to these regions should be avoided. Even among bands which are equally absorbed whatever the form of the polymer, dichroic effects may be diluted because of interaction between different vibrational modes of the same group. Zbinden (1964) has called attention to this possibility which seems not to have been considered previously in connection with polymer spectra. For instance, in the CH_2 symmetric mode of an infinite, extended polyethylene chain the transition moment would be perpendicular to the chain axis if no other motion occurred. But since there is always motion of every mode corresponding to the zero-point energy, the CH_2 group axis will have several motions, of differing frequencies, one being that corresponding to the wagging motion. This motion is slower than that of the hydrogen atoms during the CH stretching vibrations, and the transition moment for these latter is not always perpendicular to the chain axis. It is true that for motions of small amplitude the normal modes may be considered as quite independent, but for large amplitudes this approximation is perhaps not valid. Zbinden considers that the effect of other vibrations on the CH_2 stretching modes might in effect make the angle α for the CH_2 symmetric stretching vibration about 70° instead of 90°, and this leads to a dichroic ratio 2·85 (equal to the observed one) for a sample of linear polyethylene which is said to be highly oriented and whose crystallinity is said to be higher than 80%. Now a sample of polyethylene made by the high-pressure process (which is not linear, and whose crystallinity is not so high as that of linear (high density) polyethylene) has been made for which the dichroic ratio of the same band was in excess of 4·0 (Ambrose, Elliott and Temple, 1949a). The effect of the wagging mode and of other vibrations in reducing the dichroism is clearly less (and possibly a good deal less) than the 20° suggested by Zbinden. In fact, the classical concept of vibration used by Zbinden is not valid for the 'vibrations' with $v = 0$ which give rise to the zero-point energy. In a classical vibration the system spends most of its time at the extremes of the amplitude. The eigenfunction of the vibration with $v = 0$ is, however, a graph with a maximum at the mid-point of the 'vibration', and therefore the equilibrium position of the nuclei is the position in which the system spends most of its time. (The author is indebted to Dr R. D. B. Fraser for a discussion which clarified this matter.)

It is possible that an effect similar to the one described above is to be seen in Fig. 4.3. This shows dichroism in an oriented crystal layer of N-methyl acetamide (Bradbury and Elliott, 1963b; and some unpublished results). In the low-temperature form whose spectrum is shown, the unit cell is orthorhombic and the planes of all the planar molecules are parallel to *ac*. The radiation was along the *b* axis and spectra are shown with the **E** vector along *a* and along *c*. The Amide I band at 1647 cm^{-1} cannot be seen when the **E** vector is along the *c* axis and the dichroic ratio is at least 100 : 1. The peak of the Amide A band at 3270 cm^{-1} (the NH stretching mode) does show weakly with the **E** vector along *c*; this might be because the vibrations which cause in-plane deformation of the NH band (Amide II at *c*. 1580 cm^{-1}) reduce the dichroism of the NH stretch in the same way as the CH_2 deformations are said to do for the CH stretching modes. However, in N-methyl acetamide there is no reason why the transition moment of the individual NH stretching modes should be along the *a* axis. The situation is further complicated by two other factors. In the first place, the NH stretching mode is complex and has many peaks whose origin is uncertain, and further, there is an element of disorder in the crystals which may persist even at low temperatures (Bradbury and Elliott, 1963b). Spectra run at higher temperatures than those shown in Fig. 4.3 show a considerably reduced dichroism of the amide bands. This is probably because at higher temperatures the amount of disordered material is greater (there is a transition at 10° C, above which temperature the unit cell is a statistical one, as found by Katz and Post (1960), but because of impurities in the thin layers used for infra-red spectroscopy this transition is not sharp). The dichroic ratio of the NH stretching band is Fig. 4.3 is *c*. 33 : 1, which would correspond to a transition moment direction 10° out of the molecular plane. The effect just discussed causes a very small reduction in the dichroism of crystal spectra when the transition moments are approximately in a plane normal to the radiation, and it would also be small in polymer spectra when the transition moment is along or nearly along the fibre axis. However, when the inclination angle α is near 90°, the effect may be much greater.

3 Detection of 'crystallinity' bands and bands associated with amorphous polymer

It was noticed in early work with polarized infra-red radiation that although for most polymer bands the dichroic ratios were low, for some bands it was very high. One such band was found in the overtone and combination band region of the spectrum of polyethylene at 4216 cm^{-1} by Glatt and Ellis (1947) and another was found in the spectrum of oriented polyvinyl alcohol at 1146 cm^{-1} by Elliott, Ambrose and Temple (1948b). Many more examples are now known. Elliott, Ambrose and Temple showed that the high dichroism was to be expected from the high orientation of the material as shown by its x-ray diffraction pattern. It is now recognized that these bands are associated with the crystalline parts of polymers. Zerbi, Ciampbelli and Zamboni (1964) have recently considered their classification. As they point out, many bands associated with crystalline regions are a direct consequence of the regular chain conformation in the crystallites, rather than of an association of molecules in a crystal lattice. It is to be expected that crystallinity bands will show high dichroism in oriented polymers, but high dichroism alone is not sufficient to identify a band as a 'crystallinity band'.

The amorphous regions of polymers are poorly oriented, hence bands which appear only in such regions should show small dichroic effects. Again, low dichroism is not sufficient in itself to identify a band in the amorphous polymer, for the low dichroism might be a consequence of the direction of the transition moment.

More will be said about the subject of crystallinity and amorphous bands in Chapter 5.

4 Polarized radiation as an aid in the assignment of bands

A knowledge of the direction of the transition moment is almost as important as a knowledge of the frequency in assigning spectral bands to the corresponding molecular motions. For many purposes, only a distinction between parallel and perpendicular directions with respect to the fibre axis is needed and the observations need not be of great precision. There are numerous examples in the article by Krimm (1960), and in papers on isotactic polymers by Natta and co-workers, and on these and other polymers by Tadokoro and co-workers.

5 Improvement of spectral resolution

Helical polymers (Higgs, 1953) have infra-red bands whose transition moments are either parallel or perpendicular to the helix axis. When the groups forming the helix are of low symmetry, so that their individual vibrations have transition moments which are neither along nor perpendicular to the helix axis, there results a number of perpendicular and parallel bands whose frequencies are nearly the same. If unpolarized radiation is used, the two bands will rarely be resolved. With an oriented specimen and polarized radiation, resolution becomes easy. The optical density of the specimen at the peak frequency is greater for polarized than for unpolarized radiation, and this helps greatly in recognizing weak bands. A good

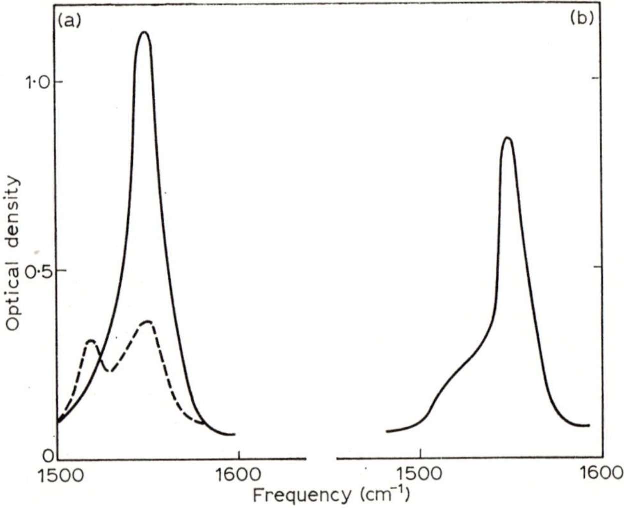

Fig. 4.13 The Amide II band in oriented poly-γ-methyl-L-glutamate. (**a**) Recorded with polarized radiation. ——, **E** vector perpendicular to fibre axis; – – –, **E** vector parallel to fibre axis. (**b**) The same spectrum as it would appear with completely unpolarized radiation.

example is seen in Fig. 4.13 which shows the spectrum of poly-γ-methyl-L-glutamate recorded with unpolarized radiation and a well-oriented sample recorded with polarized radiation. The Amide II band, as was shown by Miyazawa and Blout (1961) is split into two components in the manner explained in Chapter 5, section C, but because of the hydrogen bonds in the α-helix the separation of the components is considerable ($\sim$30 cm^{-1}). Even with this separation the weaker band at 1516 cm^{-1} is difficult to see on the spectrum of the unoriented material. Other examples of a similar kind are shown in Fig. 5.5.

CHAPTER 5

Some Special Topics in the Spectroscopy of Polymers

A HYDROGEN BONDING IN POLYMERS

Infra-red spectroscopy has for many years been a standard technique for the detection and estimation of hydrogen bonding between or within simple molecules. Hydrogen bonds are formed between (for instance) the hydrogen atom of an OH or NH group (in which the hydrogen is electrically positive) and atoms such as oxygen which have local concentrations of negative charge ('lone pair' electrons). The energy of interaction is considerably greater than that of the van der Waals forces between molecules, but much less than that of a valence bond, and in solids or in non-polar solvents is of the order of 5 kcal per mole. The bond is represented by the broken line

$$C{=}O\ldots H{-}N \quad \text{or} \quad {>}O\ldots H{-}O{<}$$

and its formation reduces the effective restoring force when the H–N or H–O bond is stretched by motion of the hydrogen atom. In consequence of this, the frequency of the corresponding vibration is reduced. The effect on the hydrogen deformation frequency is different, for the hydrogen bond makes the system more resistant to deformation, hence the deformation frequency is increased when such a bond is formed.

The effect in simple substances is usually investigated by measuring the spectrum of a range of concentrations in a non-polar solvent (if one can be found). Figure 5.1 shows the spectrum of an alcohol in carbon tetrachloride. With sufficiently low concentration of alcohol, no appreciable amount of hydrogen bonding occurs, and the OH stretching vibration appears as a sharp band at 3640 cm^{-1}. At higher concentrations of alcohol, a broad peak appears at *c.* 3400 cm^{-1}, and this is caused by OH vibrations in hydrogen-bonded molecules. With increasing concentration of alcohol, the

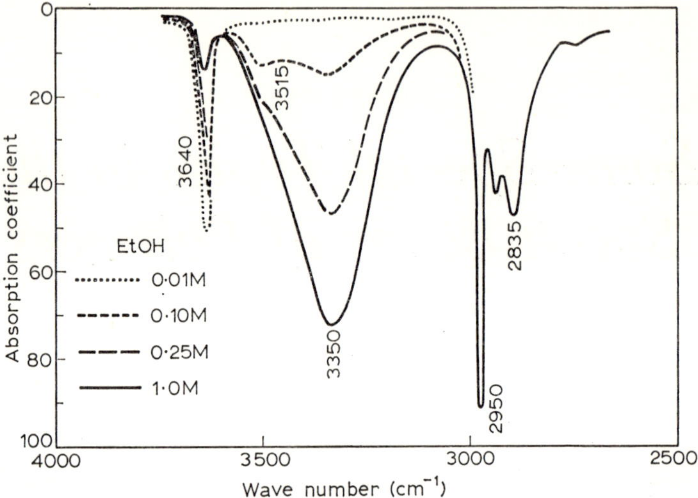

Fig. 5.1 Free and bonded OH groups. Infra-red spectra of various concentrations of ethyl alcohol in carbon tetrachloride. (Nakanishi, 1962)

height of the 'free' OH peak at 3640 cm^{-1} diminishes while that of the hydrogen-bonded OH peak increases. Similar effects can be observed with solutions of many amides, when a sharp 'free' NH band appears at 3465 cm^{-1} at low concentrations, to be replaced by the hydrogen-bonded NH band at *c.* 3300 cm^{-1} as the concentration is increased. These frequencies are found for amides in which the *trans* form

$$O{=}C{-}N{<}^{H}_{R}$$

can occur. When the amide group is in the *cis* form

$$O{=}C{-}N{<}^{R}_{H}$$

(as, for instance, in lactams), the frequencies are about 3420 cm^{-1} for the free form and *c.* 3170–3240 cm^{-1} for the bonded form (Tsuboi, 1949; see Bamford, Elliott and Hanby, 1956 for a review).

The presence of hydrogen-bonded structures in some polymers can be inferred at once from the presence of the bonded form of the hydrogen stretching mode. Thus, in polyvinyl alcohol the OH band at 3400 shows that the molecular chains form hydrogen bonds (Thompson and Torkington, 1945). In polyamides, polypeptides and proteins the hydrogen-bonded NH band at *c.* 3300 and complete absence of a 'free' NH band likewise show the formation of hydrogen bonds throughout the polymer. Some synthetic polypeptides of high molecular weight are soluble in non-polar or only slightly polar solvents (which do not form hydrogen bonds), and it has been found that in such solutions no 'free' NH band appears even at very low polymer concentrations (Ambrose and Elliott, 1951a). The explanation offered (now known to be the correct one) was that internal hydrogen bonds were formed within each polymer chain. Polypeptides in not-too-polar solvents take the α-helix form, as strongly suggested by work on the dispersion of optical rotation (Moffitt and Yang, 1956). This has recently been confirmed by x-ray diffraction (Parry and Elliott, 1966).

There appears to be a general correspondence between the lowering of the hydrogen stretching frequency and the strength of the hydrogen bond as judged by the separation of the atoms on either side (Lord and Merrifield, 1953). This relation holds from the strong H–F . . . H–F bonds (FF distance, 2·24 Å; frequency change, 2700 cm^{-1}) to very weak hydrogen bonds. However, since there are complicating factors, care should be used in interpreting very small differences in the frequency change (for instance, in different conformations of polyamide or polypeptide chains). The hydrogen-bonded OH stretching bands are very broad, and sometimes have several peaks. This happens even in the spectra of simple substances in crystalline form, and a completely satisfactory explanation is difficult to find, though partial explanations which have been offered may be correct as far as they go (see Hadži, 1959; Pimentel and McClellan, 1960). The NH stretching band in polymers has a weaker component at 3060 cm^{-1} and in some instances a still weaker one at *c.* 2820 cm^{-1}. The origin of these bands and of other fine structure is considered in the following section. They are mentioned here to emphasize that the spectra of hydrogen-bonded materials are not simple. Other references to hydrogen bonding will be found in later sections.

The subject of hydrogen bonding in solids has recently been treated in a very general way in a book by Hamilton and Ibers (1968).

B THE VIBRATIONS OF THE AMIDE (PEPTIDE) GROUP

The *trans* form of the amide group occurs in the technically important polyamides (nylons) and in the biologically important proteins. The *cis* form is of rare occurrence in polymers. Because of resonance, the CN peptide bond has partial double-bond character and the vibrational modes tend to involve all the atoms of the group

$$\begin{matrix} \alpha C \searrow & & \nearrow H \\ & C \cdots N & \\ O \nearrow & & \searrow \alpha C \end{matrix}$$

The vibrational bands of the *trans* form are as follows.

AMIDE A AND B (*c*. 3300 and 3060 cm^{-1})

These are NH stretching modes. There are two because there is an accidental resonance (known as Fermi resonance) between the NH vibration and another amide band which is probably the first overtone of the Amide II band (see below), as a result of which two separate vibrational modes occur (Badger and Pullin, 1954; Miyazawa, 1960, 1962; Cannon, 1959; Tsuboi, 1964). The Amide A and B bands both disappear when the NH groups are replaced by ND, and appear to have the same direction of transition moment. They are replaced by two bands of nearly equal intensity at *c*. 2470 and 2414 cm^{-1}. This is ascribed to Fermi resonance between ν_{ND} and the deutero forms of Amide II and III (Cannon, 1959; Pivcová, Schneider, Štokr and Jakeš, 1964). There is some evidence that in the synthetic polypeptide poly-L-alanine there is a broad region from *c*. 2700–3400 cm^{-1} over which there is absorption which might be ascribed to NH groups (Elliott, 1954). Observations on well-oriented crystal layers of acetanilide (Abbott and Elliott, 1956) and N-methyl acetamide (Bradbury and Elliott, 1963b) have shown that the NH region is very complicated. This is not to be ascribed to the number of molecules in the crystal unit cell, but to motions within the molecule. The NH absorption in N-methyl acetamide is temperature-sensitive, and many peaks are seen in the spectrum at low temperatures. It is probable that combinations are formed between the NH stretching modes and some low-frequency modes not yet identified. Volkenstein, Elyashevich and Stepanov (1950) have elaborated such an explanation for the breadth of hydrogen-bonded OH and NH modes, suggesting that the low-frequency mode is one in which the hydrogen bond itself is stretched.

AMIDE I (*c*. 1650 cm^{-1})

This is usually the strongest band in the spectrum of an amide. It has long been known to be associated with stretching of the C=O bond, and to change in frequency on the formation of a hydrogen bond (in acetanilide, for instance, the band is at 1706 cm^{-1} in very dilute solution in carbon tetrachloride, when no hydrogen bonds are formed, and at 1666 cm^{-1} in the crystal). The change of frequency 40 cm^{-1} is general. Fraser and Price (1952) suggested that Amide I is an out-of-phase vibration involving stretching both the C=O and CN bonds. According to Miyazawa, Shimanouchi and Mizushima (1958), who calculated the form of the in-plane vibrations of the amide bonds in N-methyl acetamide, there is also a large amplitude of the in-plane NH deformation.

Changes in frequency of the Amide I band with chain conformation in polymers are considered in a later section.

AMIDE II (*c.* 1550 cm^{-1})

In the Amide II mode most of the motion is in-plane deformation of the NH bond, but there is also a considerable stretching of the CN bond (Miyazawa, Shimanouchi and Mizushima, 1958). Because it is not a pure NH deformation mode, the replacement of NH by ND (although it removes the Amide II band) does not result in a lowering of the frequency by a factor of about $\sqrt{2}$ as expected from a change in mass. The band disappears, and a new band (Amide II′) appears at *c.* 1450 cm^{-1}, but the frequency change is much too small to be a simple mass effect. Fraser and Price (1952) suggested that the frequency of the NH deformation mode coincided by chance with a mode in which stretching of the CN and C=O bonds occurred. As a result of the resonance, two frequencies (Amide II and Amide III) occur. On replacement of NH by ND, this resonance is destroyed, and the ND deformation band appears near 1000 cm^{-1}, while the unperturbed stretching mode appears at 1450 cm^{-1}. In the non-hydrogen-bonded form the frequency of the Amide II band is about 40 cm^{-1} lower than when the hydrogen bonds are formed.

The intensity of the Amide II band in hydrogen-bonded forms is usually about half the intensity of Amide I.

AMIDE III (*c.* 1260–1300 cm^{-1})

The Amide III band is also one in which there is a large amplitude of NH in-plane deformation, together with CN stretching. The phase relations between these two motions are reversed in Amide III compared with Amide II (Miyazawa, Shimanouchi and Mizushima, 1958). It is usually considerably weaker than Amide II.

AMIDE IV (*c.* 620 cm^{-1})

The Amide IV band is largely due to in-plane bending of the O=C–N group, and since motion of the H atom is not involved it does not shift appreciably on deuteration.

AMIDE V (*c.* 750 cm^{-1})

The Amide V band is due to out-of-plane deformation of the NH group. On deuteration of this group, the frequency is lowered to about 500 cm^{-1}, just as expected from the increase in the mass of the vibrating atoms. The band is noticeably broad at room temperatures (Kessler and Sutherland, 1953). In solid N-methyl acetamide and in some polyamides it sharpens and the peak intensity increases when the temperature is lowered.

AMIDE VI (*c.* 600 cm^{-1})

The Amide VI band is believed to be due to an out-of-plane bending mode of the C=O groups, and does not shift appreciably on deuteration of the NH groups.

C FREQUENCY SHIFTS ASSOCIATED WITH CHANGES IN CHAIN CONFORMATION IN POLYPEPTIDES AND POLYAMIDES

1 Vibrational interactions

The spectra of polypeptides and fibrous proteins have Amide I and II bands whose frequencies are different in different chain

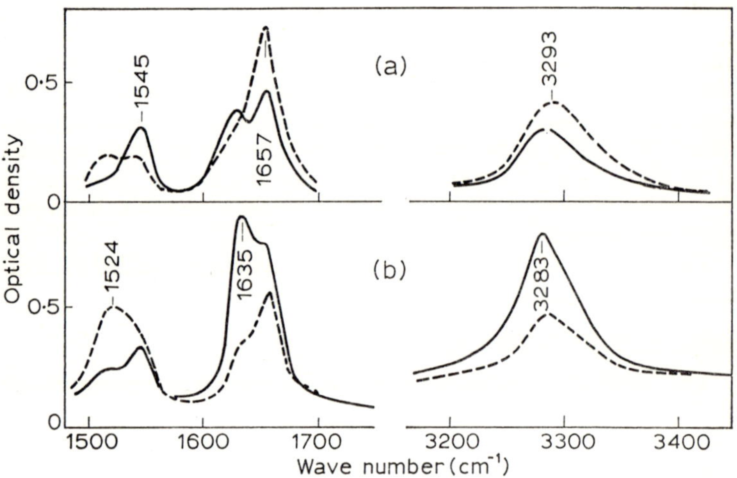

Fig. 5.2 Absorption spectrum of poly-L-alanine. (**a**) Largely α-helix form; (**b**) largely β form. ——, **E** vector perpendicular to fibre axis; – – –, **E** vector parallel to fibre axis. (Elliott, 1954)

conformations. Ambrose and Elliott (1951a) found that these frequencies are about 25–30 cm^{-1} lower for the extended form than in the α form (subsequently shown by Pauling and Corey to be the α-helix form). Figures 4.9 and 5.2 show the spectra of some synthetic polypeptides in which this can be seen. It was later found that an apparently randomly coiled form of the polypeptide chain occurred in which all the hydrogen bonds were formed, in which the frequency of the Amide I band was nearly the same as in the α-helical form, thus making it impossible to distinguish these forms by the frequency of this band (Elliott, Hanby and Malcolm, 1958). The Amide I band in the spectrum of the extended (β) form of polypeptides and proteins is a strong band of perpendicular character at *c.* 1630 cm^{-1}; it is found to be accompanied by a much weaker band at *c.* 1690 cm^{-1} of opposite character (Bamford, Elliott and Hanby, 1956). Miyazawa (1960) and Miyazawa and Blout (1961) have shown how the dependence of frequency on chain conformation follows from a consideration of the interactions and phase relations of the vibrations in amide groups which are connected by hydrogen bonds.

It is assumed that for each vibration of amide groups connected by hydrogen bonds there is a characteristic frequency ν_0 which the vibration would have were it not for the perturbing influence of the same vibrations in neighbouring amide groups. The unperturbed frequency is found in hydrogen-bonded randomly coiled forms, in which no regular phase relations between the vibration in neighbouring groups can exist. In regular forms, the interactions may be of two kinds, between neighbouring amide groups in one chain or between amide groups connected by hydrogen bonds (which need not be in one chain). The vibrational frequency, allowing for the effects of all groups is then

$$\nu(\delta, \delta') = \nu_0 + \underset{\text{intra}}{\sum_{s} D_s \cos (s\delta)} + \underset{\text{inter}}{\sum_{s'} D_s' \cos (s'\delta')} \qquad (5.1)$$

Here δ is the phase angle between the vibrations of adjacent groups and D_s is a coefficient for the interaction between the sth neighbours in the chain. The primed symbols refer to interchain, the unprimed symbols to intrachain interactions. As explained in Chapter 3, there are two infra-red-active species of vibrations in a helix, one in which the phase angle δ is zero (parallel vibrations), the other in which it is equal to ϕ, the rotation which (with a suitable translation) brings one group into coincidence with its neighbour.

2 α-helix forms

In the α-helical form of polypeptide there are 3·6 residues in one turn of the helix, hence $\delta = 0$ or $2\pi/3{\cdot}6$. Intrachain bonds are formed between every pair of third neighbours, so the interaction term D_3 will be as important as D_1. There are no interchain hydrogen bonds. The frequency of the parallel band will be

$$\nu(0) = \nu_0 + D_1 + D_3 \qquad (5.2)$$

and for the perpendicular band

$$\nu(\phi) = \nu_0 + D_1 \cos \phi + D_3 \cos 3\phi \qquad (5.3)$$

Here the symbol in parentheses indicates the phase difference between the vibrations in neighbouring groups. The separation of the parallel and perpendicular bands is small for Amide I ($\sim$2 cm^{-1})

but much larger for Amide II. They may be seen at 1550 and 1520 cm^{-1} respectively in Fig. 4.14.

3 Extended (β) forms

Extended or nearly extended polypeptide chains can form hydrogen bonds between like-directed (parallel) or unlike-directed (antiparallel) chains. The chains have 2-fold screw axes, and in the first arrangement there are two peptide groups in the unit cell of the crystal, in the second there are four. Because of the 2-fold screw axis the angle ϕ is π, and the phase angle δ is either 0 or π.

For the parallel chains, taking the interaction of nearest neighbours,

$$\nu(0, 0) = \nu_0 + D_1 + D'_1 \text{ (parallel band)} \quad (5.4)$$

$$\nu(\pi, 0) = \nu_0 - D_1 + D'_1 \text{ (perpendicular band)} \quad (5.5)$$

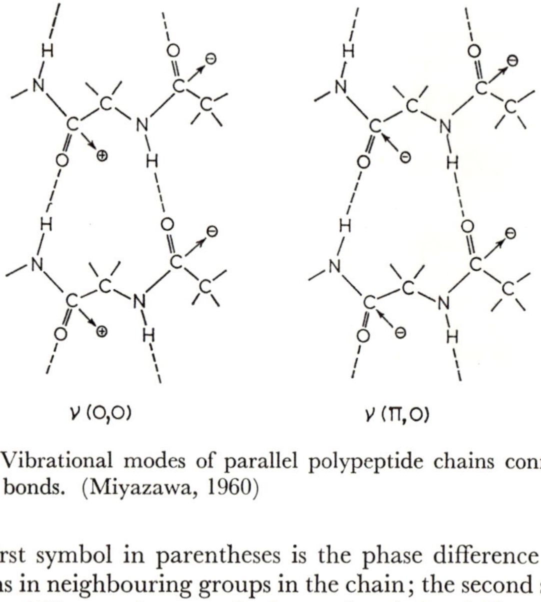

ν (0,0) ν (π,0)

Fig. 5.3 Vibrational modes of parallel polypeptide chains connected by hydrogen bonds. (Miyazawa, 1960)

The first symbol in parentheses is the phase difference between vibrations in neighbouring groups in the chain; the second symbol is the phase difference between hydrogen-bonded neighbours. Since these are in adjacent chains, and hence in adjacent unit cells, this phase difference must be zero for infra-red-active vibrations (see Chapter 3). The motions are shown in Fig. 5.3.

For the antiparallel arrangement four combinations of the phases are possible, but one of these $\nu(0, 0)$ is infra-red inactive. The other three are:

$$\nu(0, \pi) = \nu_0 + D_1 - D'_1 \text{ (parallel)} \quad (5.6)$$

$$\nu(\pi, 0) = \nu_0 - D_1 + D'_1 \text{ (perpendicular)} \quad (5.7)$$

$$\nu(\pi, \pi) = \nu_0 - D_1 - D'_1 \text{ (perpendicular)} \quad (5.8)$$

The last of these is inactive for a fully extended, planar chain and weakly active if the chain is slightly contracted. The motions are shown in Fig. 5.4.

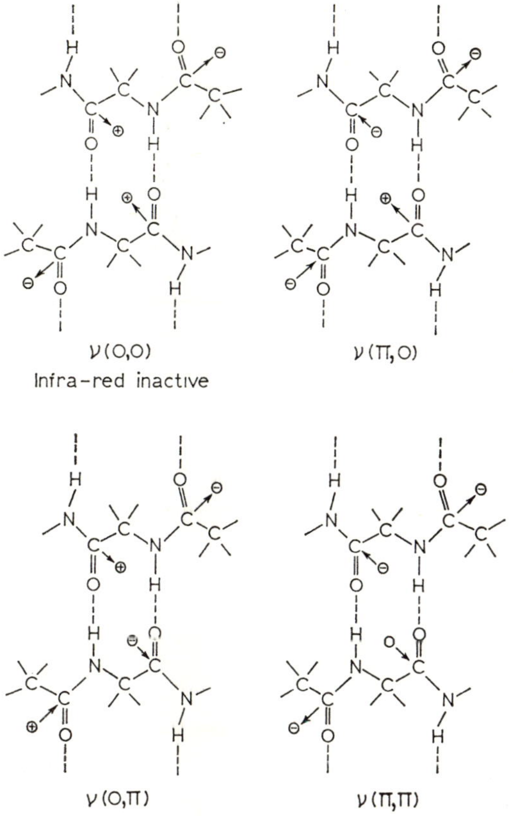

$\nu(0,0)$ Infra-red inactive

$\nu(\pi,0)$

$\nu(0,\pi)$

$\nu(\pi,\pi)$

Fig. 5.4 Vibrational modes of antiparallel chains connected by hydrogen bonds. (Miyazawa, 1960)

Miyazawa ascribed the bands seen at *c.* 1690 and 1630 cm^{-1} in the spectrum of silk and of β-synthetic polypeptides to the $\nu(0, \pi)$ and $\nu(\pi, 0)$ modes of the Amide I band in antiparallel-chain arrangements. The shift of the main Amide I band when the chain conformation is changed from an α-helix to an extended form is now seen to be better described as a greatly increased separation of the parallel and perpendicular components of the interaction. The effect of the interactions is a maximum in the antiparallel extended form, as shown by the equations above. This means that any other folded polypeptide conformations will have a frequency higher than the $\nu(\pi, 0)$ band of either the parallel or antiparallel β form. The diagnostic value of this band as an indication of the presence of β polypeptide is hereby increased.

4 Numerical values of the interaction constants ν_0, D_1 and D'_1

Miyazawa and Blout (1961) proposed certain values of the constants ν_0, D_1 and D'_1 by comparing observed frequencies of bands in

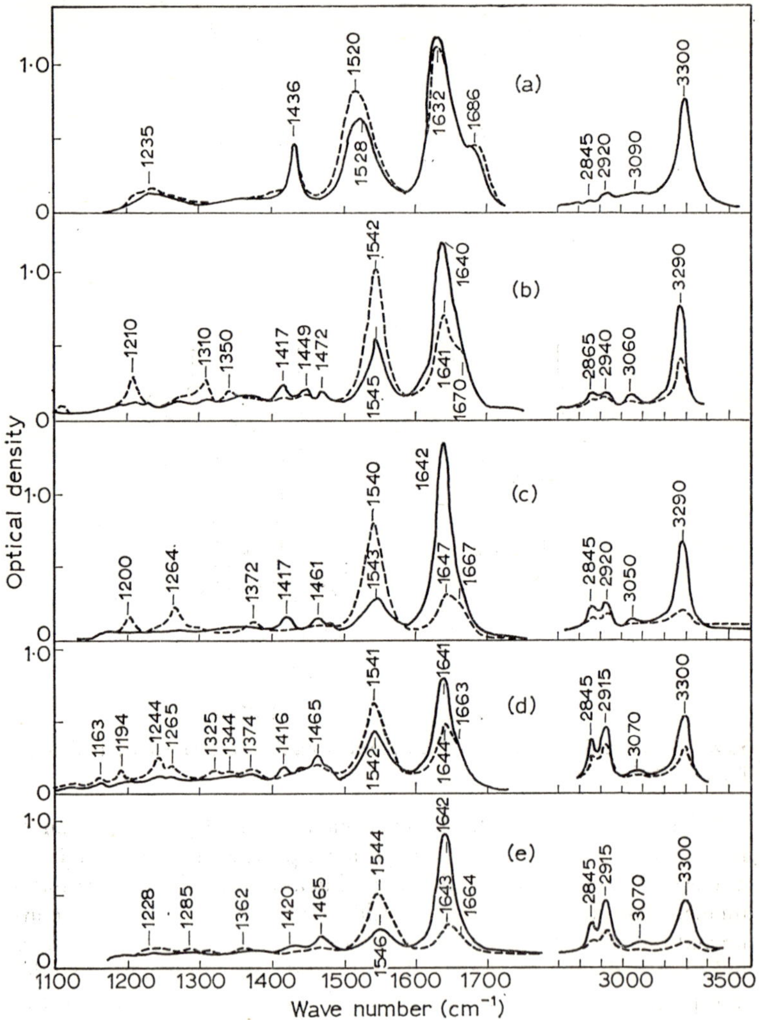

Fig. 5.5 Spectra of oriented polyamides in α form. (**a**) Nylon 2 (polyglycine)—3 μ region shown for unoriented polymer; (**b**) nylon 4; (**c**) nylon 6; (**d**) nylon 8; (**e**) nylon 10. ——, **E** vector perpendicular to fibre axis; – – –, **E** vector parallel to fibre axis. (Bradbury and Elliott, 1963a)

the spectrum of nylon 66 and of polyglycine I (the extended form). This involves the assumption that ν_0 is the same for both of these polymers. In order to avoid such an assumption, Bradbury and Elliott (1963a) determined them from a homologous series of poly-ω-amino-acids, using the members with odd numbers of CH_2 groups, nylons 2, 4, 6, 8 and 10 (the first of these is more familiar as polyglycine). These polymers all form crystalline arrangements of antiparallel chains; the higher members 6–10 also exist in the γ form in which the chains are slightly contracted but 'parallel' (see Chapter 4 for references). The spectra of these polymers, oriented in the antiparallel arrangement, is shown in Fig. 5.5. Since the polymers differ only in the length of hydrocarbon chain between the amide groups it is obvious that the interaction constant D'_1 will be the same for all, whereas D_1 will diminish rapidly with increasing distance between amide groups in one chain. In consequence of this, the separation of the $\nu(0, \pi)$ and $\nu(\pi, 0)$ bands into which Amide I is split diminishes on going to the higher members of the series, and

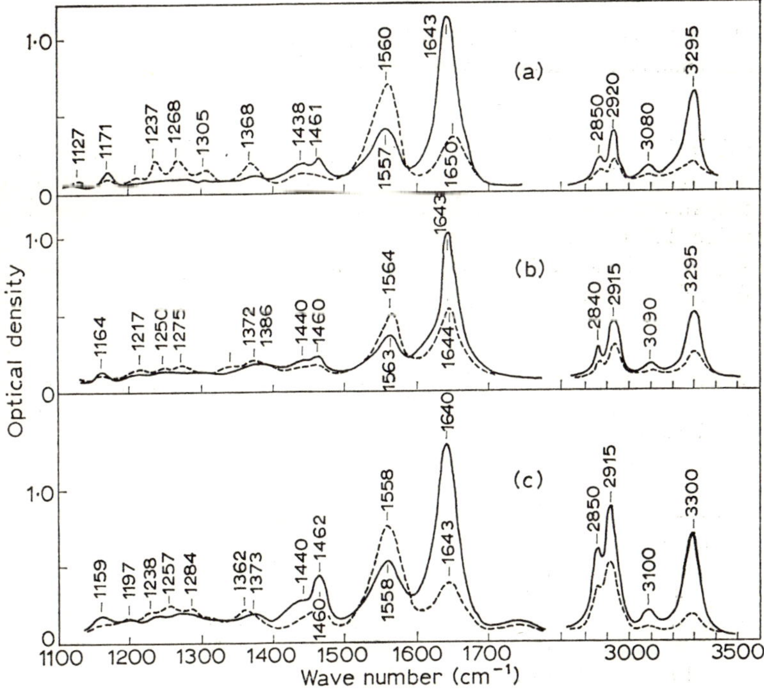

Fig. 5.6 Spectra of oriented polyamides in γ form. (**a**) Nylon 6; (**b**) nylon 8; (**c**) nylon 10. ——, **E** vector perpendicular to fibre axis; – – –, **E** vector parallel to fibre axis. (Bradbury and Elliott, 1963a)

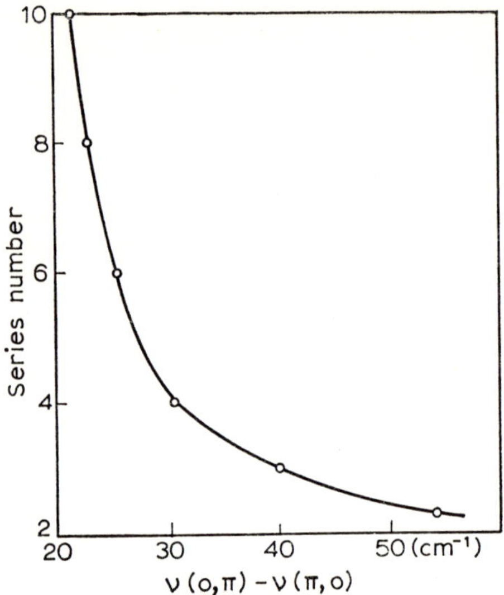

Fig. 5.7 Separation of Amide I components plotted against polyamide series number. (Bradbury and Elliott, 1963a)

this is clearly seen in Fig. 5.5. The splitting is much less in Amide II. The separation of the Amide I components is plotted against the polyamide series number in Fig. 5.7, and tends to an asymptotic value 22 cm^{-1}, which from equations (5.6) and (5.7) gives $D'_1 = -11$ cm^{-1}. The values of ν_0 and of D_1 may be calculated from the observed frequencies and are given in Table 5.1.

Table 5.1 Wave numbers of Amide I band observed in poly-ω-amino-acids (α, β form) and values of D_1 and ν_0 (cm^{-1}) (Bradbury and Elliott, 1963a)

Polymer	$\nu(\pi, 0)$	$\nu(0, \pi)$	ν_0	D_1
nylon				
2	1632	1686	1659	16
4	1640	1670	1655	4
6	1642	1667	1654½	1½
8	1641	1663	1652	0
10	1642	1664	1653	0

The intermolecular interaction constant D'_1 is taken as -11 cm^{-1} throughout.

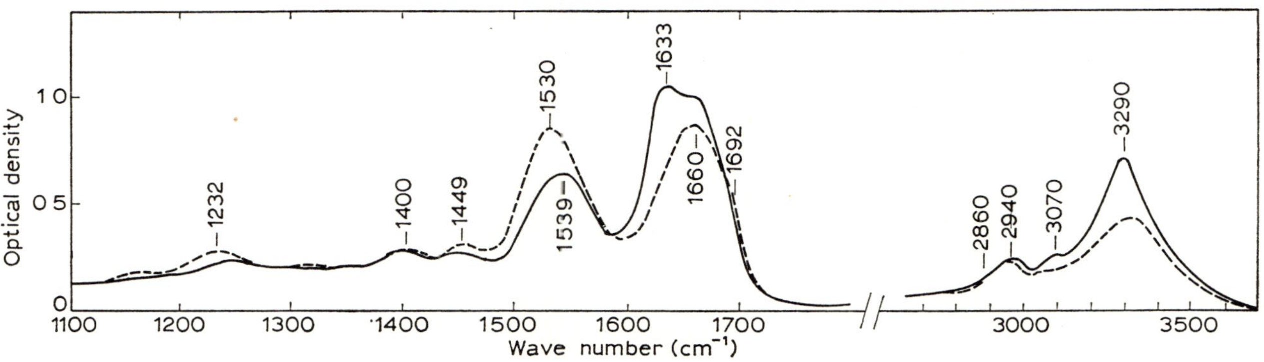

Fig. 5.8 Infra-red spectrum of steam-stretched (β) keratin (horse hair). ——, **E** vector perpendicular to fibre axis; – – –, **E** vector parallel to fibre axis. (Bradbury and Elliott, 1963a)

5 Occurrence of parallel and antiparallel chains in proteins

If the interaction constants for polyglycine given in Table 5.1 can be applied to other polypeptides, then we expect for parallel sheets to find Amide I bands at 1632 and 1664 cm^{-1}, while for antiparallel sheets the bands should appear at 1632 and 1686 cm^{-1}. In fact, the high-frequency band of the antiparallel arrangement is sometimes at least as high as 1702 cm^{-1}, and it seems likely that ν_0 is somewhat variable from polymer to polymer. Many β-polypeptide spectra

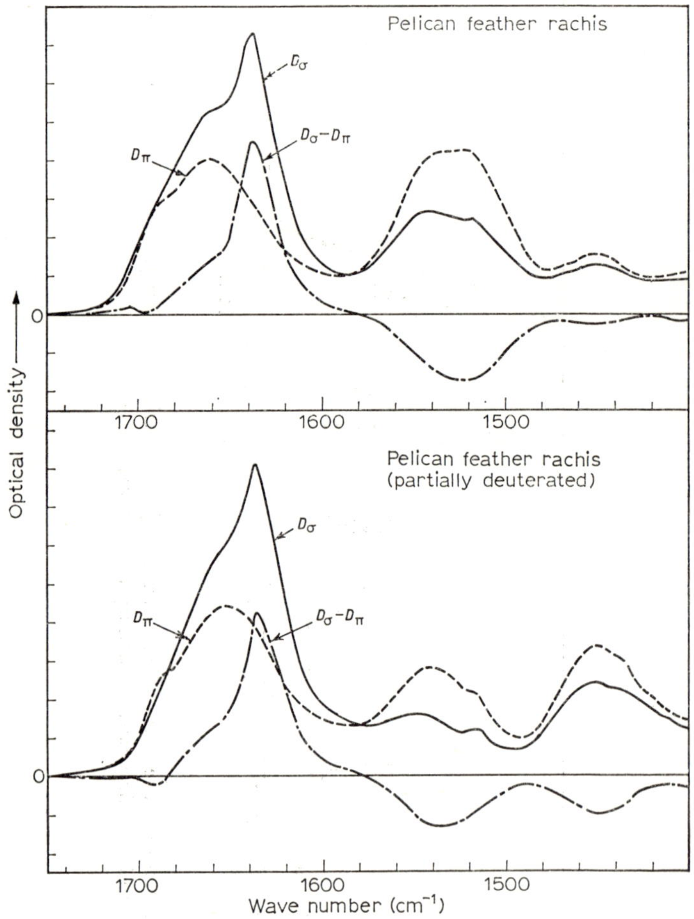

Fig. 5.9 Infra-red spectrum of natural (β) feather keratin. ——, **E** vector perpendicular to fibre axis; – – –, **E** vector parallel to fibre axis. (Fraser and Suzuki, 1965b)

have been found with the $\nu(0, \pi)$ band, and the antiparallel arrangement seems to be common. It has been observed in the steam-stretched β form of hair keratin as shown in Figs. 5.8 and 5.14 (Bradbury and Elliott, 1963a; Bendit, 1966) and also (Fig. 5.9) in the β-keratin of pelican feather (Fraser and Suzuki, 1965b). These proteins also have a band at *c.* 1660 cm^{-1} which appears both with perpendicular and with parallel orientation of the **E** vector with respect to the fibre axis, and this (since it is broad) could conceal a band at *c.* 1664 cm^{-1} from a 'parallel' arrangement of molecular chains. The broad band probably indicates a random coil arrangement.

Although feather keratin has been described as a β-protein for many years, the interpretation of the x-ray diffraction pattern is not straightforward, and the molecules appear to be formed into small units which are arranged in a helical manner (Fraser and MacRae, 1963). The spectroscopic evidence that the structure of the molecular chains is really of the near-extended, β type is therefore of some importance.

The spectrum of a cut section of silk gut (from *Bombyx mori*) shows the parallel $\nu(0, \pi)$ and perpendicular $\nu(\pi,0)$ bands very clearly (Fig. 4.7). This material is known from x-ray analysis to be in the antiparallel extended form (Marsh, Pauling and Corey, 1955). The spectrum in the region 1600–1700 cm^{-1} has been investigated in some detail by Suzuki (1967), who separated the components of the spectra (obtained with polarized radiation) by computation. One of the parameters which was varied to obtain the best fit was the band half-width; the $\nu(0, \pi)$ band was found to be rather less than half as broad as the $\nu(\pi,0)$ band. In consequence of this, it is necessary to use integrated intensities instead of peak intensities in order to calculate the transition moment direction of the Amide I band. When this is done, the transition moment for an unperturbed vibration is found to make an angle 19° with the C=O bond. This is in good agreement with values found in crystalline compounds (see Table 4.1). The difference in the half-widths of the $\nu(0, \pi)$ and $\nu(\pi, 0)$ bands may be connected with the shape of the crystallites in silk fibroin, which are elongated in the direction of the fibre axis.

The analysis showed a weakly dichroic (though strong) band of frequency 1656 cm^{-1} which is ascribed to the random phase. This material amounts to about 70% of the whole, only 30% being crystalline.

One spectrum has been recorded in which the $\nu(0, 0)$ band of a parallel chain arrangement of polypeptide chains appears. This is a weak band of frequency 1676 cm^{-1} which appears in the cross-β form of poly-β-*n*-propyl-L-aspartate (Bradbury and Elliott, 1963a). This polymer appears to be mainly in the antiparallel arrangement (as suggested by the x-ray diffraction pattern as well as by the occurrence of a band at 1702 cm^{-1}). There appear to be faults in the antiparallel sheet, however, so from place to place there occur two (or more) like-directed chains in proximity, giving indications of a parallel-chain component. The dichroism of the 1676 cm^{-1} band is of the perpendicular character which would be expected for the $\nu(0, 0)$ band in a cross-β form.

The infra-red spectra and structure of polypeptides and proteins form the subject of a book by Chirgadze (1965).

6 Frequency changes in nylon 6 (polycaprolactam)

As described in Chapter 4, two well-defined forms of nylon 6 are known and the structure of these is known from x-ray diffraction. The extended, planar form (known as α, or A) forms monoclinic crystallites with antiparallel chains connected by hydrogen bonds (Holmes, Bunn and Smith, 1955). The γ (sometimes called B) form is often described as hexagonal or pseudo-hexagonal though it is more accurately orthorhombic (Bradbury, Brown, Elliott and Parry, 1965) and has tilted amide groups; the hydrogen bonds are formed between parallel chains (Kinoshita, 1959). In the monoclinic α form the $\nu(\pi, 0)$ and $\nu(0, \pi)$ bands of Amide I are well separated because of the antiparallel arrangement, and may be seen in Fig. 5.5. In the γ form, the separation of the $\nu(0, 0)$ and $\nu(\pi, 0)$ according to equations (5.4) and (5.5) depends on the term D_1, which is very small because the interacting amide groups are separated by five methylene groups. The Amide I band of the γ form appears single in consequence (Fig. 5.6). This distinction between the α and γ forms is, however, not of any analytical use, though it is important in understanding the relation between polymer structure and spectrum. Other, more striking differences may be seen in Figs. 5.5 and 5.6. Schneider, Doskočilová and co-workers have made extensive studies of the structure and spectra of nylon 6 and other polyamides, and have found that nylon 6 gives a third spectrum, different from that of the α or γ forms, which they ascribe to amorphous polymer. This spectrum is obtained from a melt of nylon 6 at 245° C or from a molten nylon 6 sample quenched in liquid nitrogen, and observed at temperatures below −65° C. The spectrum of amorphous nylon 6 has far fewer peaks than that of the α or γ forms, and no resolved bands are seen below 550 cm^{-1}, a region in which skeletal bending modes of straight-chain paraffins appear. From this it appears that the polymethylene chain between the amide groups is not straight in the amorphous form. The same conclusion is drawn from the absence of the complicated band structure between 1200 and 1300 cm^{-1} in the spectrum of the α and γ forms. This structure is caused in part by coupling of methylene twisting and wagging vibrations with Amide III. Some bands are found to be characteristic of the three forms, others are independent of chain conformation and so their intensity may be used as a measure of the thickness of the nylon 6 sample. Doskočilová, Pivcová, Schneider and Čifelin (1963) find that the monoclinic α form is characterized by bands at 575, 692, 930, 960, 1201, 1420 and 1477 cm^{-1}. The pseudo-hexagonal γ form has characteristic bands at 625 and 711 cm^{-1}, but these bands also occur in a mesomorphic, less ordered state which has the same chain conformation. The amorphous polymer has absorption bands at 590, 700, 1240 and 1280 cm^{-1}, but these bands overlap those of other forms. Bands which are common to all forms occur at 730, 1075, 1120, 1370, 2860 and 2930 cm^{-1}. The bands at 976, 1170, 1440 and 1460 cm^{-1} are characteristic of all types of polymethylene chains not coplanar with the amide group. The bands at 440, 520 and 1266 cm^{-1} are characteristic of the straight polymethylene chain, irrespective of the orientation of the amide groups. With a detailed knowledge of the relation between band frequency and structure it becomes possible to use spectroscopic methods to investigate the changes in polymer structure produced by heat-treatment, mechanical working, etc. More detailed assignments of the absorption bands to atomic motions have been given for the α form of nylon 6 by Jakeš, Schmidt and Schneider (1964).

D THE SUBSTITUTION OF HYDROGEN BY DEUTERIUM ATOMS

The substitution of an atom by one of its isotopes leaves the potential field unchanged, so the frequency of molecular vibrations is affected only by the change in mass. Vibrations in which most of the motion is in the hydrogen atoms in a polymer chain will show a large decrease in frequency when these atoms are replaced by deuterium. CH, NH and OH stretching modes are of this kind, because they are usually far removed in frequency from other molecular modes. If the vibration were restricted to the hydrogen atoms and the atoms to which they are bonded, the frequency would depend on the inverse of the square root of the reduced mass. If M and H are the masses of the two atoms in the hydrogenated form of the molecule, and M and D those in the deutero form, the reduced masses are, respectively, $MH/(M + H)$ and $MD/(M + D)$. The frequency is therefore reduced by a factor on substituting deuterium for hydrogen. When M is carbon, nitrogen or oxygen this factor is near 1·37. Because other atoms also take part in the motion to some extent, the factor is a little lower, and variable. It is usually near 1·33. Similar frequency changes may be observed in deformation modes, but not infrequently these modes are coupled to other modes of similar frequency in which deuterium substitution has a much smaller effect, and so the frequency change may be much less than that expected. Vibrational modes in which, for instance, a whole CH_2 group moves together (wagging and twisting modes) will be little affected by deuteration.

Deuterium substitution is a powerful technique for checking the assignment of bands to the appropriate molecular vibration. The replacement of the various species of CH groups by CD, separately as well as together, is a common procedure in polymer study. The substitution once made, the CD groups are stable and do not readily revert to CH. The hydrogen atoms attached to oxygen and nitrogen atoms are much more labile. Replacement by deuterium in these cases is usually done by exposing the polymer to D_2O vapour, or by dissolving and casting the polymer from a solvent which has labile hydrogen atoms which have been replaced by deuterium. Such solvents are heavy water, and deuterated acids such as CH_3COOD, $CHCl_2COOD$, CF_3COOD. The casting of films from these liquids must be done out of contact with atmospheric water vapour (see Chapter 2).

Various deutero analogues of polyvinyl chloride have been prepared by Krimm, Folt, Shipman and Berens (1963), and their spectra have been examined to aid the more detailed analysis of the spectrum of the hydrogenated form of the polymer.

An example of the use of deuterium substitution of polyvinyl alcohol is shown in Fig. 5.10. The polymer has the structure

$$-\left(\begin{array}{c}CH_2\text{-}CH\\ \quad\quad | \\ \quad\quad OH\end{array}\right)_n-$$

and in the usual (atactic) form the oxygen atoms may be attached to either of the carbon valence bonds not in the chain, randomly. The hydrogen of the OH groups was exchanged for deuterium by exposing to D_2O vapour, but other methods (described by Tadokoro, Nagai, Seki and Nitta, 1961) were used to replace the CH groups. In the fully hydrogenated form (PVA) the very strong, broad, non-dichroic band at 3400 cm^{-1} is interpreted as a hydrogen-bonded OH stretching mode, and this is confirmed by the spectrum of PVA–d, in which this band is quite weak (the substitution is difficult to carry to completion) and a strong, broad, non-dichroic band appears at *c.* 2500 cm^{-1}. The frequency change (ratio 1·36) is just as expected. There is a similar broad band in PVA at *c.* 1430 cm^{-1} (really a parallel band at 1440 and a perpendicular band at 1420 cm^{-1}) which might be an OH deformation mode. In PVA–d, the broad band disappears and there remains a very sharp band at the same frequency, which is probably a CH_2 deformation mode. In

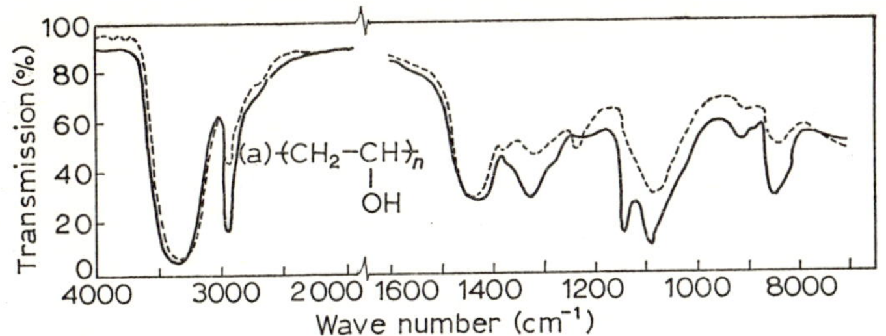

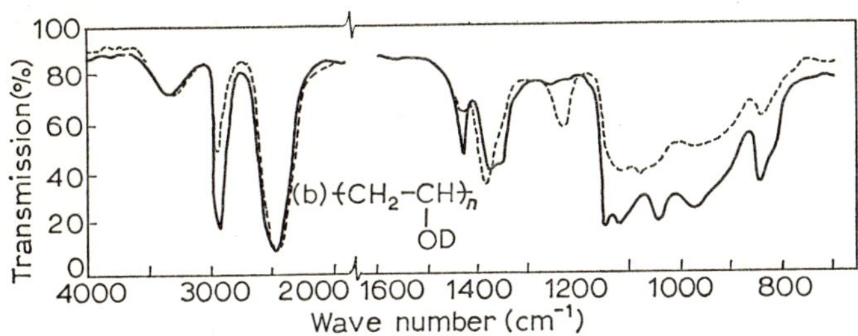

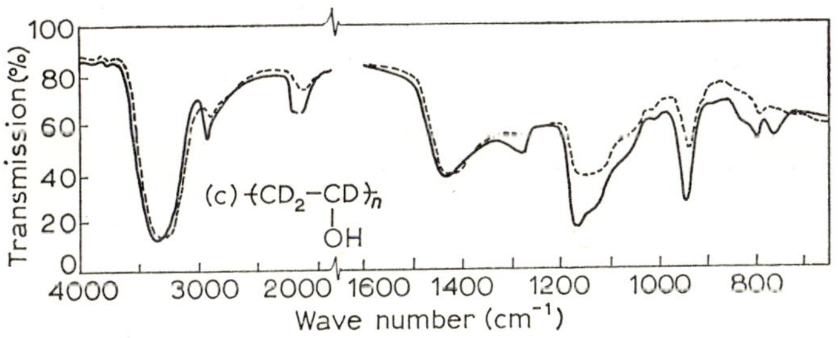

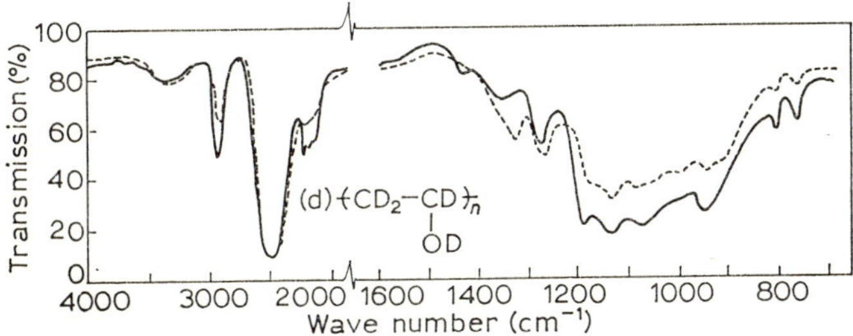

Fig. 5.10 Infra-red spectra of hydrogenic and deutero forms of polyvinyl alcohol: (**a**) PVA; (**b**) PVA–d; (**c**) PV–d_3A; (**d**) PV–d_3A–d. ——, **E** vector perpendicular to fibre axis; – – –, **E** vector parallel to fibre axis. (Tadokoro, Nagai, Seki and Nitta, 1961)

the hydrogenated form these two vibration modes may be coupled. It will be seen that the broad, almost non-dichroic band appears in PV–d_3A, in which the CD_2 deformation mode will have a lower frequency (it cannot be identified with certainty in the region of unresolved bands between 1000 and 1200 cm^{-1}). The CH_2 stretching modes remain unaffected by the deuteration of the OH groups, but the corresponding CD_2 bands appear at 2200 cm^{-1} in PV–d_3A and PV–d_3A–d. The symmetric and asymmetric stretching modes are not resolved. It will be seen that the replacement of CH by CD is very incomplete. This can complicate the spectrum, because in a polymer many molecular species occur in these circumstances, in which the symmetry is quite different from what it would be in the pure form. Sometimes this is done deliberately, a specimen being prepared in which one isotopic species is present in a very small amount. This allows the vibrations of isolated parts of a molecule to be examined. It has been used in crystal spectra, and Miyazawa (1960) has used it to determine the frequency of Amide II when there is hydrogen bonding but no permanent phase relation with surrounding amide groups. This frequency is the ν_0 referred to in section C on frequency changes associated with chain conformation. It was observed in poly-γ-benzyl-L-glutamate in which most of the NH groups were replaced by ND. The residual Amide II band of the hydrogenated form appeared at 1535 cm^{-1}, very near the mean frequency of the two bands at 1546 and 1516 cm^{-1} which appear in the hydrogenated form of the polymer.

Another use of the replacement of labile hydrogen atoms by deuterium is described in the following section. It depends on the fact that if this exchange is done, for instance, by D_2O vapour, the replacement occurs more rapidly in those regions of the polymer to which the vapour has easy access. Because of this, it is possible to demonstrate and to investigate non-homogeneity in polymers, which is related to the presence of crystalline and amorphous regions.

DEUTERIUM EXCHANGE IN CELLULOSE

It was found by Frilette, Hanle and Mark (1948) that the density of cellulose in which the OH groups had been allowed to exchange with D_2O increased rapidly during the initial exchange reaction, but almost stopped after some hours. They suggested that the part which had not reacted was the more highly ordered, crystalline form which was not easily penetrated by the molecules of D_2O. Mann and Marrinan (1956) used infra-red spectroscopy to follow the reaction, by observing the diminution of the OH stretching band and the increase of the corresponding OD band. Some of the results obtained with regenerated cellulose (viscose) are shown in Fig. 5.11.

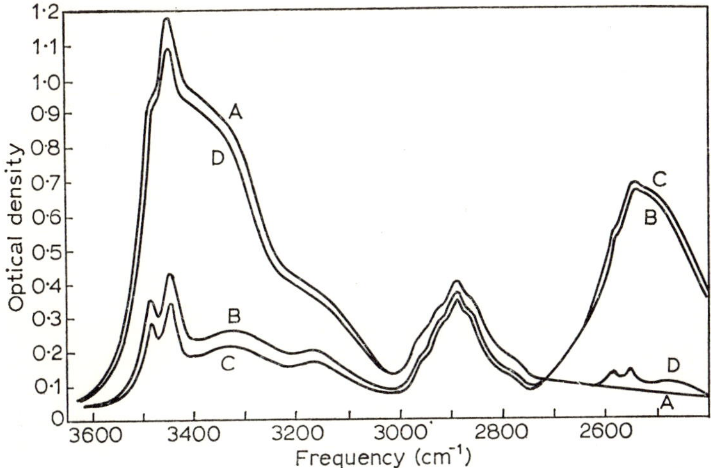

Fig. 5.11 Viscose film—exchange with D_2O. A, dry viscose undeuterated; B, deuterated in D_2O vapour for 4 hr; C, deuterated in D_2O liquid for 4 hr; D, rehydrogenated in H_2O liquid for 4 hr. (Mann and Marrinan, 1956)

The very broad OH band is rapidly reduced, giving a broad, apparently structureless OD band. After about 4 hours, the OH band shows several peaks, and even on immersion in liquid D_2O most of this structure persists. When rehydrogenation is allowed to take place, the same pattern of peaks is seen in the OD region after

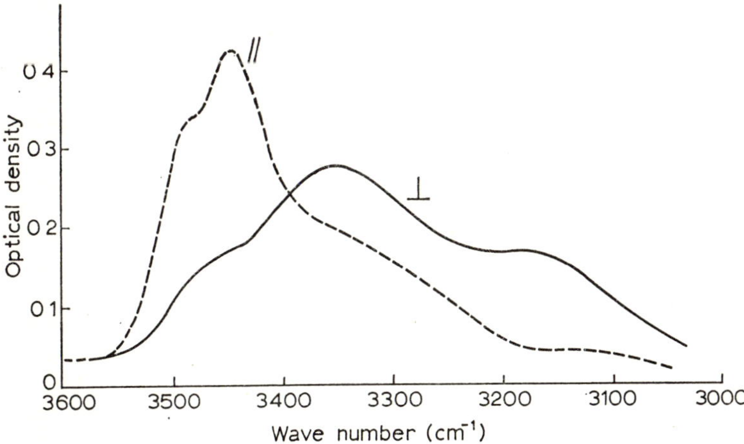

Fig. 5.12 Dichroism in crystalline regions of cellulose II. (Mann and Marrinan, 1958)

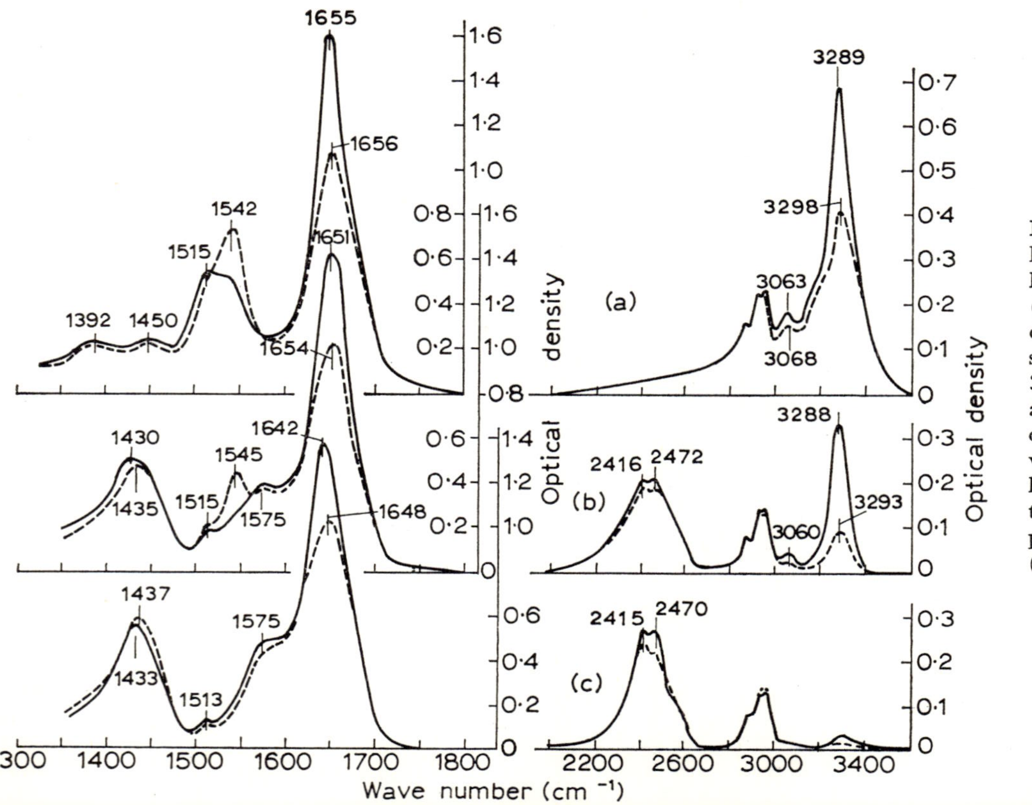

Fig. 5.13 Spectrum of α-keratin, 3·5-μ section of horse hair, at 0% R.H.: (**a**) untreated; (**b**) partially deuterated, exposed to saturated D_2O vapour at 35° C for 100 min; (**c**) almost fully deuterated, exposed to saturated D_2O vapour at 90–95° C for 4 hr. ——, **E** vector parallel to fibre axis; – – –, **E** vector perpendicular to fibre axis. (Bendit, 1966)

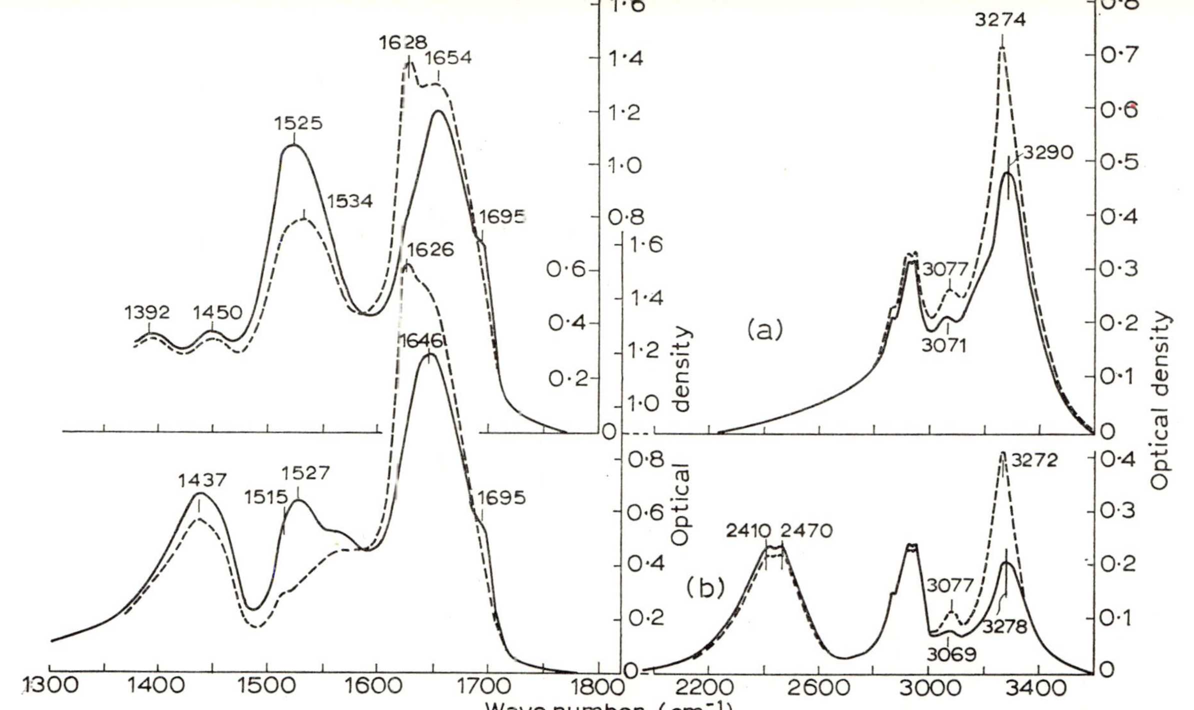

Fig. 5.14 Spectrum of β-keratin, 5-μ section of steam-stretched horse hair: (**a**) untreated; (**b**) partially deuterated, exposed to saturated D_2O vapour at 35° C for 70 min. ——, **E** vector parallel to fibre axis; – – –, **E** vector perpendicular to fibre axis. (Bendit, 1966)

most of the broad OD band has disappeared. This pattern is apparently characteristic of the more ordered parts. Dichroic measurements of the residual OH bands of viscose are shown in Fig. 5.12 (Mann and Marrinan, 1958). The same pattern is found in the residual OD bands, whether the deuteration is complete or partial, and each OH peak is lowered by the same factor on deuteration. It seems likely that the peaks correspond to different hydrogen bonds (each cellulose residue has three OH groups). From the fact that peaks with parallel as well as perpendicular dichroism are observed, it appears that some OH bonds are in the direction of the chain axis, and these may be of the intra-molecular kind.

DEUTERIUM EXCHANGE IN FIBROUS PROTEINS

Deuteration of proteins such as silk and α-keratin (hair) shows the same rapid initial rate followed by a much slower rate as is observed with cellulose, the NH stretching band being replaced by the ND band. As the NH groups in the less ordered parts are replaced, the dichroism of the residual NH band increases considerably. This has been observed in silk (Sutherland, 1955), in porcupine quill by Parker (1955) and in hair by Bendit (1966). These experiments show very clearly the presence of components of different order in these materials, though some part of the increase in dichroism is due (particularly in keratin) to the removal of bands from OH and NH_2 groups in the side chains. The effects are shown in Fig. 5.13 for natural hair, and in Fig. 5.14 for the β-keratin of steam-stretched hair.

E DIFFERENCES BETWEEN THE SPECTRA OF CRYSTALLINE AND AMORPHOUS POLYMERS

From what has been said in Chapter 3, we may expect differences between the spectra of crystalline and amorphous polymers for the following reasons.

(1) The chain conformation in crystallites will be regular (fully extended, or folded in some regular way), and in most cases only one stereo-isomer will be present. In disordered regions, many stereo-isomers will be present, and these will have vibrational frequencies, some of which are characteristic of each isomer. For instance, *gauche* forms may have different spectra from *trans* forms (see below, polyethylene terephthalate).

(2) The loss of chain symmetry in disordered forms may result in some vibrations which are not active in the chain conformation present in the crystal becoming weakly active in the amorphous part.

(3) Bands, forbidden in the isolated, regular chain conformation may become weakly active in the crystal if the symmetry of the single chain is lowered in the crystal. Such bands will not appear in disordered forms.

(4) If there are two or more chains in the unit cell of the crystal, single vibrational modes of the isolated, regular chain may become multiple in the crystalline form.

(5) Because the crystalline parts of polymers generally become more highly oriented than the amorphous parts when the polymer is stretched or rolled, dichroic effects are usually much more pronounced in bands which are associated with crystalline parts.

It is clear that a number of effects will reveal the presence of states of different order in polymers, but a clear distinction between crystalline and amorphous parts (if such exist) is difficult to make. Effects (3) and (4), if their existence can be satisfactorily demonstrated, might be held to indicate the presence of a crystalline arrangement, and such bands might well be called 'crystallinity bands'. However, many of the bands observed in the spectra of polymers which give x-ray diffraction patterns of crystalline materials are not of this kind, but belong to a particular chain conformation which is found in the crystallites, and would appear even if the chains in such a conformation were not arranged in any particular way with respect to their neighbours. Zerbi, Ciampbelli and Zamboni (1964) have proposed the name 'regularity bands' for these bands. They give as an example a set of bands which are present in the spectrum of isotactic (stereo-regular) polypropylene in the smectic or paracrystalline modification. This form is obtained by quenching (from the melt) and apparently all the chains in it have a regular helical structure, but there is no lateral order sufficient to permit the system to be called crystalline. The bands disappear when the regularity is destroyed. This can be done (*a*) by changing the geometry of the chain by melting, or (*b*) by changing the nature of some atoms in the chain by making a deuteropropylene/propylene copolymer. In this form the geometry of the helix is not expected to alter, but the masses of some atoms have changed. Spectra of polypropylene are shown in Figs. 5.15 and 5.16.

Bands which show abnormally high dichroism (5) may be associated with crystal forms, but here again this is by no means certain, for the regular form of the chain might be highly oriented without the formation of crystallites.

EXAMPLES OF BANDS SENSITIVE TO THE STATE OF THE POLYMER

The spectrum of isotactic polypropylene has many bands which are strong when the polymer is solid, but which disappear in the

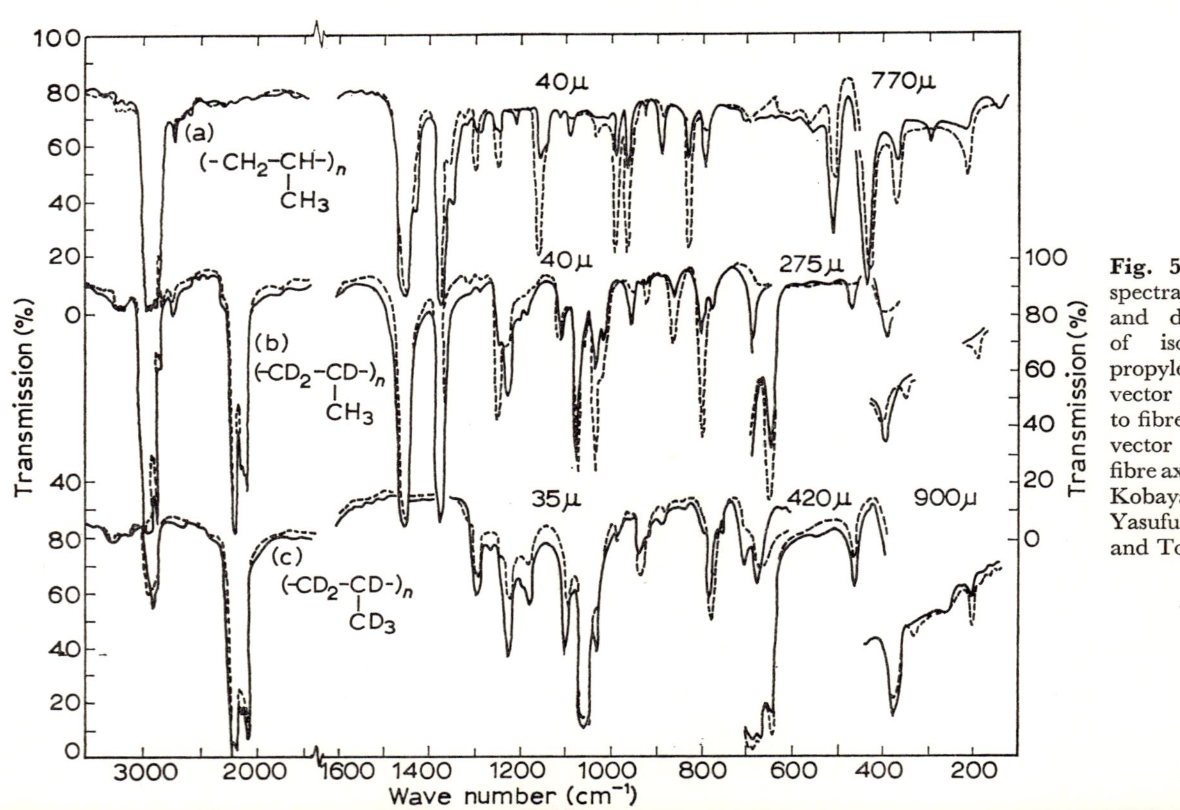

Fig. 5.15 Infra-red spectra of hydrogenic and deutero forms of isotactic polypropylene. ——, **E** vector perpendicular to fibre axis; – – –, **E** vector parallel to fibre axis. (Tadokoro, Kobayashi, Ukita, Yasufuku, Murahashi and Torii, 1965)

molten form (Fig. 5.16). They are probably 'regularity bands'. A detailed examination of the normal modes of the polymer has been made by Miyazawa (1964) and by Schachtschneider and Snyder (1964). Many of the modes are a complex mixture of internal vibrations, and upon deuteration, the form of many modes is changed so drastically that the analogous band cannot be identified in the deuterated sample.

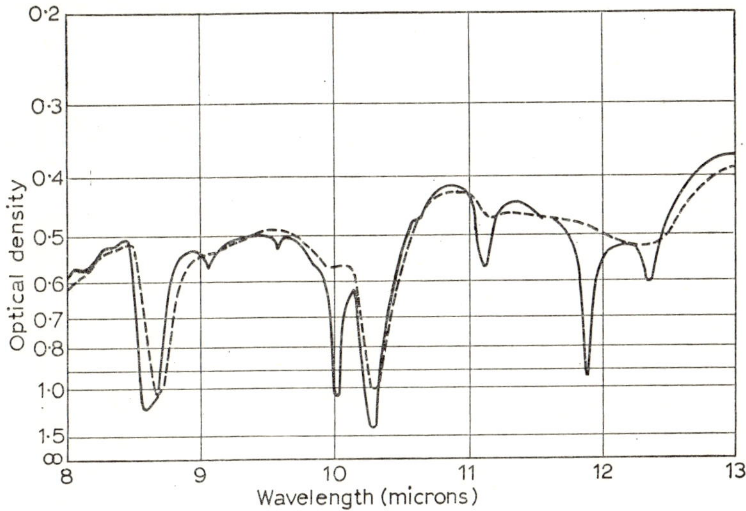

Fig. 5.16 Infra-red spectrum of isotactic polypropylene. ——, film 0·076 mm thick at room temperature; – – –, the same film molten between two salt plates at 240° C. (Zbinden, 1964)

The spectrum of polyvinyl alcohol (Fig. 5.10) has a band at 1140 cm^{-1} which is seen only in the record made when the **E** vector of the radiation is perpendicular to the direction of stretching. This may be a 'crystallinity band' but, since the polymer whose spectrum is shown was atactic, some randomness is inevitable and it may be doubted whether the band is caused by the effects (3) and (4) described above.

In polyamides, many instances may be found of bands whose intensities change greatly when the polymer is heat-treated, and for nylon 6 many of these bands have been classified (see section C of this chapter). They are mainly 'regularity bands', although the components of the Amide I bands which appear in antiparallel-chain arrangements could properly be called 'crystallinity bands'.

The spectrum of polyethylene terephthelate shows some bands which are strong in the amorphous form (obtained by rapid cooling of the molten polymer) but which are greatly weakened when the material is crystallized (Fig. 5.17). These bands have been used by Miller and Willis (1956) to measure the amorphous content of the polymer. They appear, however, to be associated with the *cis* form of the $-OCH_2CH_2O-$ part of the chain, which in the crystalline polymer is known from x-ray diffraction studies (Daubeny, Bunn and Brown, 1954) to be in the *trans* form. Although the bands at 898 and 1040 cm^{-1} must therefore come from polymer in the amorphous state, not all the amorphous polymer will absorb at these frequencies.

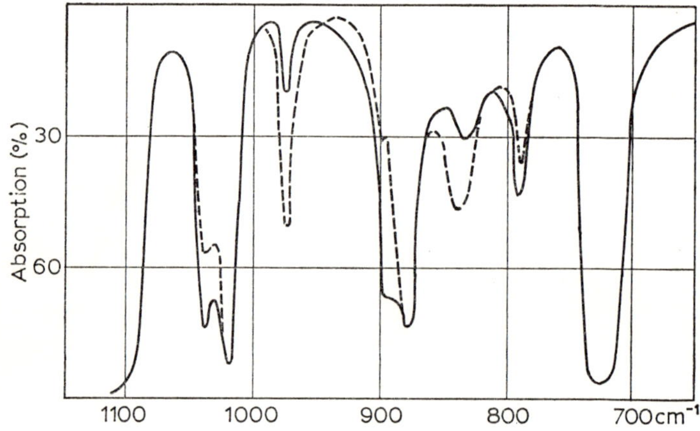

Fig. 5.17 Infra-red spectrum of polyethylene terephthalate. ——, totally amorphous; – – –, fairly crystalline. (Miller and Willis, 1956)

CHARACTERIZATION OF POLYMERS FROM INFRA-RED SPECTRA

It will be obvious that quantitative measurements of bands which are sensitive to the form of the polymer can be used to characterize the polymer, even though there may be some doubt as to whether it is really 'crystallinity', 'regularity' or 'amorphous content' which is being measured.

The strength of an absorption band is measured by the optical density at its peak. If the incident and transmitted intensities of the radiation of frequency ν are respectively I_0 and I, the optical density is

$$\log_{10} (I_0/I) = k_\nu lC \tag{5.9}$$

where k_ν is the extinction coefficient per mole fraction per unit length, C is the concentration of absorbing material expressed as a mole fraction, and l is the length of the absorbing path. This relation is true if Beer's law holds.

If a polymer could be obtained wholly in one form, then for a band which appears only in this form C would be unity in such a specimen, and from measurement of the optical density and thickness k_ν could be found. Similar measurements on other specimens would enable the concentration C of the form in question to be determined. It is often possible to obtain wholly amorphous specimens, and this procedure can be used for such materials. For 'crystallinity bands', however, a different procedure is necessary, since wholly crystalline high polymers cannot usually be obtained, and k_ν must be found from a specimen whose crystallinity has been determined by other methods. Density measurements have often been used to give an independent estimate of crystallinity (although in some polymers the presence of voids makes this method inaccurate). If d_c is the density of the pure crystalline form (which can be obtained from x-ray measurements of the unit cell), d_l the density of the amorphous (liquid) phase, the density d of the polymer in bulk is related to the fraction of crystalline material C_c by

$$C_c = (d - d_l)/(d_c - d_l). \tag{5.10}$$

The need for accurate thickness measurements can be avoided if a band can be found whose intensity does not depend on the form of the polymer. If Beer's law holds, for the sensitive band (be it 'crystalline', 'regularity' or 'amorphous' band)

$$\log_{10} (I_0/I)_s = k_s C_s l \tag{5.11}$$

For the independent band $C = 1$, and

$$\log_{10} (I_0/I)_i = k_i l \tag{5.12}$$

Hence

$$C_s = \frac{k_i \log_{10} (I_0/I)_s}{k_s \log_{10} (I_0/I)_i} \tag{5.13}$$

By making measurements on a specimen in which C_s is known, the ratio k_i/k_s can be determined, and subsequent measurements on other specimens allow the calculation of the fraction of the form corresponding to the sensitive band.

References

ABBOTT, N. B. and ELLIOTT, A. (1956) *Proc. R. Soc.* A, **234**, 247.
AMBROSE, E. J. and ELLIOTT, A. (1951a) *Proc. R. Soc.* A, **205**, 47.
AMBROSE, E. J. and ELLIOTT, A. (1951b) *Proc. R. Soc.* A, **206**, 206.
AMBROSE, E. J. and ELLIOTT, A. (1951c) *Proc. R. Soc.* A, **208**, 75.
AMBROSE, E. J., ELLIOTT, A. and TEMPLE, R. B. (1949a) *Proc. R. Soc.* A, **199**, 183.
AMBROSE, E. J., ELLIOTT, A. and TEMPLE, R. B. (1949b) *Nature, Lond.*, **163**, 859.
AMBROSE, E. J., ELLIOTT, A. and TEMPLE, R. B. (1951) *Proc. R. Soc.* A, **206**, 192.
AMBROSE, E. J. and HANBY, W. E. (1949) *Nature, Lond.*, **163**, 483.
ARIMOTO, H. (1964) *J. Polym. Sci.* A, **2**, 2283.
ASTBURY, W. T., DICKINSON, S. and BAILEY, K. (1935) *Biochem. J.*, **29**, 2351.
BADGER, R. M. and PULLIN, A. D. E. (1954) *J. chem. Phys.*, **22**, 1142.
BAMFORD, C. H., ELLIOTT, A. and HANBY, W. E. (1956) *Synthetic Polypeptides.* Academic Press, New York and London.
BAMFORD, C. H., HANBY, W. E. and HAPPEY, F. (1951a) *Proc. R. Soc.* A, **205**, 30.
BAMFORD, C. H., HANBY, W. E. and HAPPEY, F. (1951b) *Proc. R. Soc.* A, **206**, 407.
BAUMAN, R. P. (1962) *Absorption Spectroscopy.* John Wiley, New York.
BEER, M. (1956) *Proc. R. Soc.* A, **236**, 136.
BELLAMY, L. J. (1958) *The Infra-red Spectra of Complex Molecules*, 2nd edn., Methuen, London.
BENDIT, E. G. (1966) *Biopolymers*, **4**, 561.
BRADBURY, E. M., BROWN, L., DOWNIE, A. R., ELLIOTT, A., FRASER, R. D. B. and HANBY, W. E. (1962) *J. molec. Biol.*, **5**, 230.
BRADBURY, E. M., BROWN, L., DOWNIE, A. R., ELLIOTT, A., FRASER, R. D. B., HANBY, W. E. and MCDONALD, T. R. R. (1960) *J. molec. Biol.*, **2**, 276.
BRADBURY, E. M., BROWN, L., ELLIOTT, A. and PARRY, D. A. D. (1965) *Polymer*, **6**, 465.
BRADBURY, E. M. and ELLIOTT, A. (1962) *J. scient. Instrum.*, **39**, 390.
BRADBURY, E. M. and ELLIOTT, A. (1963a) *Polymer*, **4**, 47.
BRADBURY, E. M. and ELLIOTT, A. (1963b) *Spectrochim. Acta*, **19**, 995.
BRADBURY, E. M. and FORD, M. A. (1966) *Appl. Optics*, **5**, 235.
BUNN, C. W. (1939) *Trans. Faraday Soc.*, **35**, 482.
CANNON, C. G. (1959) in *Physical Methods of Investigating Textiles*, Eds. R. MEREDITH and J. W. S. HEARLE. Textile Publishers Inc., New York; Interscience, Chichester.
CHARNEY, E. (1955) *J. opt. Soc. Am.*, **45**, 980.
CHIRGADZE, YU. N. (1965) *Infrared Spectra and Structure of Polypeptides and Proteins*. Hayka, Moscow (in Russian).
COATES, V. J., OFFNER, A. and SIEGLER, E. H. (1953) *J. opt. Soc. Am.*, **43**, 984.
COLTHUP, N. B. (1950) *J. opt. Soc. Am.*, **40**, 397.
CONN, G. K. T. and EATON, G. K. (1954) *J. opt. Soc. Am.*, **44**, 553.

DAUBENY, R. de P., BUNN, C. W. and BROWN, C. J. (1954) *Proc. R. Soc.* A, **226** 531.
DOSKOČILOVÁ, D., PIVCOVÁ, H., SCHNEIDER, B. and ČIFELIN, P. (1963) *Collection Czech. chem. Commun.*, **28**, 1867.
ELLIOTT, A. (1952) *Proc. R. Soc.* A, **211**, 490.
ELLIOTT, A. (1953) *Proc. R. Soc.* A, **221**, 104.
ELLIOTT, A. (1954) *Proc. R. Soc.* A, **226**, 408.
ELLIOTT, A. (1959) in *Advances in Spectroscopy*, vol. 1, Ed. H. W. THOMPSON. Interscience, New York and London.
ELLIOTT, A. and AMBROSE, E. J. (1947) *Nature, Lond.*, **159**, 641.
ELLIOTT, A., AMBROSE, E. J. and TEMPLE, R. B. (1948a) *J. opt. Soc. Am.*, **38**, 212.
ELLIOTT, A., AMBROSE, E. J. and TEMPLE, R. B. (1948b) *J. chem. Phys.*, **16**, 877.
ELLIOTT, A., HANBY, W. E. and MALCOLM, B. R. (1958) *Discuss. Faraday Soc.*, **25**, 167.
FORD, M. A., PRICE, W. C., SEEDS, W. E. and WILKINSON, G. R. (1958) *J. Opt. Soc. Amer.*, **48**, 249.
FAHRENFORT, J. (1961) *Spectrochim. Acta*, **17**, 698.
FRASER, R. D. B. (1950) *Discuss. Faraday Soc.*, **9**, 378.
FRASER, R. D. B. (1953a) *J. opt. Soc. Am.*, **43**, 929.
FRASER, R. D. B. (1953b) *J. chem. Phys.*, **21**, 1511.
FRASER, R. D. B. (1956) *J. chem. Phys.*, **24**, 89.
FRASER, R. D. B. (1958a) *J. chem. Phys.*, **28**, 1113.
FRASER, R. D. B. (1958b) *J. chem. Phys.*, **29**, 1428.
FRASER, R. D. B. (1960) in *A Laboratory Manual of Analytical Methods of Protein Chemistry*, vol. 2, Eds. P. ALEXANDER and R. J. BLOCK. Pergamon, Oxford.
FRASER, R. D. B. and MACRAE, T. P. (1963) *J. molec. Biol.*, **7**, 272.
FRASER, R. D. B. and PRICE, W. C. (1952) *Nature, Lond.*, **170**, 490.
FRASER, R. D. B. and SUZUKI, E. (1965a) *Spectrochim. Acta*, **21**, 615.
FRASER, R. D. B. and SUZUKI, E. (1965b) *J. molec. Biol.*, **14**, 279.
FRASER, R. D. B. and SUZUKI, E. (1966) *Analyt. Chem.*, **38,** 1770.
FRILETTE, V. J., HANLE, J. and MARK, H. (1948) *J. Am. chem. Soc.*, **70**, 1107.
GEBBIE, H. A. and CANNON, C. G. (1952) *J. opt. Soc. Am.*, **42**, 277.
GLATT, L. and ELLIS, J. W. (1947) *J. chem. Phys.* **15**, 884.
HADŽI, D. (1959) in *Hydrogen Bonding*, Eds. D. HADŽI and H. W. THOMPSON. Pergamon Press, Oxford.
HAMILTON, W. C. and IBERS, J. A. (1968) *Hydrogen Bonding in Solids*. Benjamin, New York and Amsterdam.
HARRICK, N. J. (1967) *Internal Reflection Spectroscopy*. Interscience, New York.
HEARLE, J. W. S. (1963) in *Fibre Structure* Eds. J. W. S HEARLE and R. H. PETERS. Butterworth, London.
HENNIKER, C. J. (1967) *Infrared Spectroscopy of Industrial Polymers*. Academic Press, London and New York.
HERZBERG, G. (1945) *Infra-red and Raman Spectra of Polyatomic Molecules*. van, Nostrand, New York; Macmillan, London.
HIGGS, P. W. (1953) *Proc. R. Soc.* A, **220**, 472.
HOLLIDAY, P. (1949) *Nature, Lond.*, **163**, 602.
HOLMES, D. R., BUNN, C. W. and SMITH, D. J. (1955) *J. Polym. Sci.*, **17**, 159.
HOSEMANN, R. (1962) *Polymer*, **3**, 349.
HVIDT, A. and LINDERSTRØM-LANG, K. (1955) *Biochim. biophys. Acta*, **16**, 168.
JAKEŠ, J., SCHMIDT, P. and SCHNEIDER, B. (1965) *Collection Czech. chem. Commun.*, 30, 996.
JONES, R. N. (1952) *J. Am. chem. Soc.*, **74**, 2681.
KATZ, J. L. and POST, B. (1960) *Acta crystallogr.*, **13**, 624.
KESSLER, H. K. and SUTHERLAND, G. B. B. M. (1953) *J. chem. Phys.*, **21**, 570.
KINOSHITA, Y. (1959) *Makromolek. Chem.*, **33**, 1.

KRIMM, S. (1954) *J. chem. Phys.*, **22**, 567.

KRIMM, S. (1960) *Fortschr. HochpolymForsch.*, **2**, 51.

KRIMM, S., FOLT, V. L., SHIPMAN, J. J. and BERENS, A. R. (1963) *J. Polym. Sci.*, **A1**, 2621.

KRIMM, S., LIANG, C. Y. and SUTHERLAND, G. B. B. M. (1956) *J. chem. Phys.*, **25**, 549.

LIANG, C. Y. and KRIMM, S. (1956) *J. chem. Phys.*, **25**, 563.

LORD, R. C. and MERRIFIELD, R. E. (1953) *J. chem. Phys.*, **21**, 166.

MacGILLAVRY, C. H. (1941) *Rec. Trav. chim.*, **60**, 605.

MAKAS, A. S. and SHURCLIFF, W. A. (1955) *J. opt. Soc. Am.*, **45**, 998.

MANN, J. and MARRINAN, H. J. (1956) *Trans. Faraday Soc.*, **52**, 481, 487, 492.

MANN, J. and MARRINAN, H. J. (1958) *J. Polym. Sci.*, **32**, 357.

MARSH, R. E., COREY, R. B. and PAULING, L. (1955) *Biochim. biophys. Acta*, **16**, 1.

MATSUI, Y., KUBOTA, T., TADOKORO, H. and YOSHIHARA, T. (1965) *J. Polym. Sci.* A, **3**, 2275.

MILLER, R. G. J. and WILLIS, H. A. (1956) *J. Polym. Sci.*, **19**, 485.

MILLER, R. G. J. and WILLIS, H. A. (1961) in *Molecular Spectroscopy*. Part II, Eds. G. H. BEAVEN, etc. Heywood, London.

MITSUI Y., IITAKA, Y. and TSUBOI, M. (1967) *J. molec. Biol.* **24**, 15.

MIYAZAWA, T. (1960) *J. chem. Phys.*, **32**, 1647.

MIYAZAWA, T. (1962) in *Polyamino Acids*, Ed. M. A. STAHMANN. University of Wisconsin, Madison.

MIYAZAWA, T. (1964) *J. Polym. Sci.* C, **7**, 59.

MIYAZAWA, T. and BLOUT, E. R. (1961) *J. Am. chem. Soc.*, **83**, 712.

MIYAZAWA, T., SHIMANOUCHI, T. and MIZUSHIMA, S. (1958) *J. chem. Phys.*, **29**, 611.

MOFFITT, W. and YANG, J. T. (1956) *Proc. natn. Acad. Sci. U.S.A.*, **42**, 596.

MORRISON, J. D. and ROBERTSON, J. M. (1949) *J. chem. Soc.*, p. 987.

NAKINISHI, K. (1962) *Infra-red Absorption Spectroscopy*. Nankodo Co. Ltd., Tokyo.

NEWMAN, R. and HALFORD, R. S. (1948) *Rev. scient. Instrum.*, **19**, 270.

NIELSEN, J. R. (1964) *J. Polym. Sci.* C, **7**, 19.

NIELSEN, J.R. and HOLLAND, R. F. (1961) *J. molec. Spectrosc.*, **6**, 394.

NIELSEN, J. R. and WOOLLETT, A. H. (1957) *J. chem. Phys.*, **26**, 1391.

NORRIS, K. P. (1954) *J. scient. Instrum.*, **31**, 284.

NORRIS, K. P., SEEDS, W. E. and WILKINS, M. H. F. (1951) *J. opt. Soc. Am.*, **41**, 111.

PARKER, K. D. (1955) *Biochim biphys. Acta*, **17**, 148.

PARKER, K. D. and RUDALL, K. M. (1957) *Nature Lond.*, **179**, 905.

PARRY, D. A. D. and ELLIOTT, A. (1966) *Nature Lond.*, **206**, 616.

PAULING, L. and COREY, R. B. (1951) *Proc. natn. Acad. Sci. U.S.A.*, **37**, 241.

PIMENTEL, G. C. and MCCLELLAN, A. L. (1960) *The Hydrogen Bond*. Freeman, San Francisco and London.

PITHA, J. and JONES, R. N. (1966) *Can. J. Chem.*, **44**, 3031.

PIVCOVÁ, H., SCHNEIDER, B., ŠTOKR, J. and JAKEŠ, J. (1964) *Collection Czech. chem. Commun.*, **29**, 2436.

PFUND, A. G. (1906) *Johns Hopkins Univ. Circ.*, No. 4, 13.

POTTS, W. J. Jr. (1963) *Chemical Infrared Spectroscopy* vol I. John Wiley, New York and Chichester.

RAMACHANDRAN, G. N. and KARTHA, G. (1955) *Nature Lond.*, **176**, 593.

ROBINSON, D. Z. (1951) *Analyt. Chem.*, **23**, 273.

ROBINSON, T. S. and PRICE, W. C. (1953) *Proc. phys. Soc.* B, **66**, 969.

SANDEMAN, I. (1955) *Proc. R. Soc.* A, **232**, 105.

SCHACHTSCHNEIDER, J. H. and SNYDER, R. G (1964) *J. Polym. Sci.* C, **7**, 99.

SCHAEFFER, C. and MATOSSI, F. (1928) *Das Ultrarote Spektrum.* Julius Springer, Berlin.

SNYDER, R. G. (1957) *J. chem. Phys.*, **27**, 969.

SNYDER, R. G. (1960) *J. molec. Spectrosc.*, **4**, 411.

SNYDER, R. G. (1967a) *J. molec. Spectrosc.*, **23**, 224.

SNYDER, R. G. (1967b) *J. chem. Phys.*, **47**, 1316.

SUTHERLAND, G. B. B. M. (1955) *Rend. Ist. Lombardo Sci.*, **89**, 67.

SUZUKI, E. (1967) *Spectrochim. Acta*, **23A**, 2303.

TADOKORO, H. (1960) *J. chem. Phys.*, **33**, 1558.

TADOKORO, H., KOBAYASHI, M., UKITA, M., YASUFUKU, K., MURAHASHI, S. and TORII, T. (1965) *J. chem. Phys.*, **42**, 1432.

TADOKORO, H., NAGAI, H., SEKI, S. and NITTA, I. (1961) *Bull. chem. Soc. Japan*, **34**, 1504.

TASUMI, M. and SHIMANOUCHI, T. (1965) *J. chem. Phys.*, **43**, 1245.

TASUMI, M. and KRIMM, S. (1967) *J. chem. Phys.*, **46**, 755.

TASUMI, M. and KRIMM, S. (1968) *J. Polym. Sci.*, **A2**, **6**, 995.

THOMPSON, H. W. (1960) Tables of wave numbers for the calibration of infra-red spectrometers (Commission of the International Union of Pure and Applied Chemistry). *J. pure appl. Chem.*, **1**, 603.

THOMPSON, H. W. and TORKINGTON, P. (1945) *Proc. R. Soc.* A, **184**, 3.

TSUBOI, M. (1949) *Bull. chem. Soc. Japan*, **22**, 215 and 255.

TSUBOI, M. (1962) *J. Polym. Sci.*, **59**, 139.

TSUBOI, M. (1964) *Biopolym. Symp.*, **1**, 527.

UEDA, S. and KIMURA, T. (1958) *Chemy high Polym., Tokyo*, **15**, 243.

VOGELSONG, D. C. (1963) *J. Polym. Sci.* A, **1**, 1055.

VOLKENSTEIN, M., ELYASHEVICH, M. A. and STEPANOV, B. I. (1950) *Zh. fiz. Khim.*, **24**, 1153.

WALSH, A. (1952) *J. opt. Soc. Am.*, **42**, 94 and 96.

WRIGHT, N. J. (1948) *J. opt. Soc. Am.*, **38**, 69.

ZBINDEN, R. (1964) *Infrared Spectroscopy of High Polymers.* Academic, New York and London.

ZERBI, G., CIAMPBELLI, F. and ZAMBONI, V. (1964) *J. Polym. Sci.* C, **7**, 141.

Index